THE GENERATOR

HOW MATHEMATICIANS DON'T THINK

RICHARD SPIERS

Order this book online at www.trafford.com/07-1181
or email orders@trafford.com

Most Trafford titles are also available at major online book retailers.

Note for Librarians: A cataloguing record for this book is available from Library
and Archives Canada at www.collectionscanada.ca/amicus/index-e.html

ISBN: 978-1-4251-3194-4

*We at Trafford believe that it is the responsibility of us all, as both individuals and corporations,
to make choices that are environmentally and socially sound. You, in turn, are supporting this
responsible conduct each time you purchase a Trafford book, or make use of our publishing services.
To find out how you are helping, please visit www.trafford.com/responsiblepublishing.html*

*Our mission is to efficiently provide the world's finest, most comprehensive book publishing
service, enabling every author to experience success. To find out how to publish your book, your
way, and have it available worldwide, visit us online at www.trafford.com/10510*

 www.trafford.com

North America & international
toll-free: 1 888 232 4444 (USA & Canada)
phone: 250 383 6864 ♦ fax: 250 383 6804 ♦ email: info@trafford.com

The United Kingdom & Europe
phone: +44 (0)1865 722 113 ♦ local rate: 0845 230 9601
facsimile: +44 (0)1865 722 868 ♦ email: info.uk@trafford.com

10 9 8 7 6 5 4 3 2 1

For Muriel

Cover Photograph: 'Koala Waking'

Sculpture and photograph by Richard Spiers

I made this sculpture in 2004.
It is on the cover because I like the subject
And the sculpture.

An animal who sleeps for a large part of his life might
Have a connection with what is revealed in this book.

INTRODUCTION

I discovered a series which produces all possible Pythagorean triples. First step is to check for myself that it is a discovery. I did that and became convinced that what I had found really was new.

The method of calculating each value in the series is very simple. The discovery lay in seeing how these triples exist in the x^2 series. And that is not complicated because the x^2 series is not complicated. It hasn't got any hiding places. There needs to be a small but significant shift in thinking and then anyone with an interest in the subject can see what is there and come to that, always has been there.

Next step is obvious - take it to a mathematician and ask him to look at it.

"Mr. Mathematician, would you please look at what I have found? Am I right in thinking it is a discovery? If so is it likely to be of any interest or importance?"

A polite and reasonable approach, and the mathematician himself is both interested and polite enough to see me. One would expect that to be the tenor of what ensued.

But it wasn't. Not one bit. I was mauled by an attack dog of mathematics and it was a most unpleasant experience.

So the reader might well conclude that I had come up with a load of rubbish. But that was not the case. Just getting past the nit-picking savagery to actually present what I had found was very, very difficult. Once that point was reached, were faults then found? No. Was the series shown to be already known? No.

I tried with other mathematicians and got the same result. The attacks were characterised by an unreasoning desire not to look at what I had found. When this could no longer be avoided, the attack dogs managed to combine two contradictory attitudes: an apparent incomprehension and the absolute certainty that what was being shown was totally unimportant.

I am not now, nor have I ever been, a mathematician. I won't ever become one either. And the culture of mathematics does not allow anyone who is not a mathematician to have a mathematical thought. Any originality of thought or discovery from outside the discipline is forbidden.

I began to think deeper, not just about the series I had found but why mathematics did not know of it. Certainly, there could be no issue of complexity beyond the reach of all but a very few minds. If I could find this series, then almost anyone could. A shift of thinking? Yes, it does need that. Complex mathematical workings? Definitely not.

What follows in this book is a description of the series that I found and some of the discoveries and insights I gained from this. When I started to relate my discovery to the culture of mathematics I was surprised at just how much evidence there is to support my view that this, the most rigorous discipline ever devised by man, has a severely restrictive culture. The real loser from that culture is the discipline itself, and those who need a flexible, imaginitive, and open-minded mathematics - and that is all of us.

I give only the barest essentials of the mathematics involved in the discovery. There is sufficient data to show that the series exists, and its basic elements. The attack dogs are getting some stripped bones and a reasonable bowl of dry biscuits. The only meat I could serve them would be very raw mathematical meat and it seems that this is what drives them crazy, so I've stripped out as much of it as I can.

For those of us who have struggled with a subject which so often seemed unnecessarily difficult, even arcane, I hope that what follows will shed some light.

CHAPTER 1

DIVIDING THE SQUARE

All good mathematics should start with a definition – or should it? I'd rather show where a Pythagorean triple exists and how it gets to be there. If anyone is stuck on what a triple is, then read on and you'll probably pick it up while you go.

Take the number 91; square it, and then divide by 7.

$$91^2 \ = \ 8281$$
$$8281 \div 7 \ = \ 1183$$

That means that there are seven units, each of 1183, that make up 91^2, and now I'm going to write each one across the page:

1183	1183	1183	1183	1183	1183	1183

Could those seven equal values be recast as a sequence of seven consecutive odd numbers that would still total to 8281? Obviously, *"But why?"* Indulge me.

1177	1179	1181	1183	1185	1187	1189

The sequence of all odd numbers is the difference between the sequence of all square numbers. The bit of the x^2 series containing these numbers looks like this:

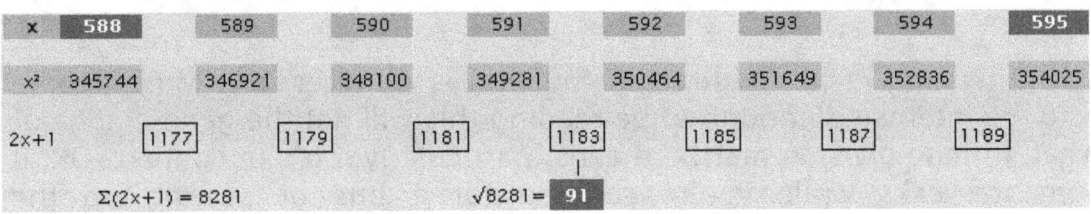

x	588	589	590	591	592	593	594	595
x^2	345744	346921	348100	349281	350464	351649	352836	354025
2x+1	1177	1179	1181	1183	1185	1187	1189	

$\Sigma(2x+1) = 8281$ $\sqrt{8281} =$ 91

The highlighted numbers are the Pythagorean triple 91,588,595. It is one of those three-part ratios that obeys the theorem we all learned – *the square on the hypotenuse equals the sum of the squares on the other two sides.*

If one triple can be identified in this manner in the x^2 series, then all triples can be so identified.

The Triples Generator is an infinite division matrix which produces all Pythagorean triples. To do so it uses all of the counting numbers as divisors and makes all of the counting numbers from 3 onwards available as dividend values. Any triple is identified by its three numbers, not by the square values, and all three must be whole numbers for it to be a triple. The triple shown above is the sixth in the Division by 7 series of the generator.

The attack dogs are already snarling. If any reader has a real need, right now, to know what this all might be then I suggest you turn to 'Bones and Biscuits' further on in the book. Otherwise, carry on here and quite a lot will become apparent.

The phrase 'infinite division matrix' earned me a particular mauling. It seems that what fired the dogs was that they had never heard the term before. Once I had recovered from the attack I took some heart. If mathematics did know of the series, then no matter who had found it and what it was called, the description would end up much the same:

- 'infinite' – having no end.
- 'division' – it is one of the four arithmetic operators; its use in the description would mean that dividing is likely to be important in this thing.
- 'matrix' – having two axes to form a grid. If division is going on, one would be for the divisors, the other for the dividends.

All that I've just written in 'bullets' is unnecessary. Most people who are interested enough to be reading this will get the general idea of what 'infinite division matrix' means, particularly if it's in context. OK, if someone asks while you're running for a bus or serving up the spaghetti it might not register, but the dogs were well and truly in context.

The real problem was the culture they were trying to serve. In reacting as they did against something that was obviously mathematical, put forward by an obviously non-mathematical person, it became clear that 'attack' was all they knew or could conceive. Evaluation was not within the mind-set.

In 'Bones and Biscuits' I have shown at Table 1 the start of The Triples Generator, and the pages of Table 2 show how some of those early triples can be found in the x^2 series. It is not a requirement that each triple generated has to be physically placed in the x^2 series, but they can be. Normally they would be listed in tabular form and a few of them are shown in Tables 3 and 4.

The value of placing a triple in the x^2 series is to give a simple and graphic understanding of how the x^2 series works. It starts at zero and squares every number in sequence thereafter. The key is the sequence of odd numbers being the difference between the sequence of squares. Starting with the first of these, 1, and keep adding then the point at which you get fed up, hopefully quite early, you will have the next square value.

What the generator does is to divide the square of each divisor's dividend values and find the place in this line of differences where there is a continuous set of odd numbers that equals the divided square. The quantity of differences used is bound to be the same as the divisor's own value. That's what happened in the example at the start of this chapter.

Wherever that set of odd numbers is, there must be a square at the beginning and another one at the end. And that is the triple: a set of odd numbers equal to the square of the first value, then the square at the start of the set and finally, the square at the end of it.

OK, the triple itself is shown as the numbers, not their square values but, hey, this isn't rocket science and that interested reader I spoke of earlier isn't going to go into an apopletic incomprehension about the numbers rather than the square values.

CHAPTER 2

DIVISION BY 1 AND THE SHIFT OF THINKING

Multiplication is speed addition. Forming the square of a number is a multiplying process. But a square value can be just as easily formed by adding the sequence of odd numbers. Admittedly, one needs to keep a free idea of what 'easily' means. The job of adding all the odd numbers to get, say, 109561 (331^2) gives watching paint dry a good name. It isn't a question of what humans see as 'easy'; we bring a whole load of stuff that's got nothing to do with the task itself and in this case boredom is very high up the list. 'Easily' means not having any intrinsic difficulty. But what we humans bring of our emotions when facing any task – that is a concept worth bearing in mind.

The square of any odd number is another odd number. So the square of every odd number must stand as a single value in the line of differences. There's a square before it, a square after it, and that's a triple. There is an infinity of them and they have the form 5,12,13 or 51,1300,1301. With only one difference lying between the other two squares, the second and third numbers of the triple must be consecutive. The early triples of this type are shown as they arise in the x^2 series in Table 2.1. A slightly longer tabular listing is provided in Table 3.1.

This business of the square of an odd number being a single value of difference is very well known. I started by showing a triple which is produced by Division by 7. Working back down from there, the stop position is going to be Division by 1. That isn't really a good reason to have this idea of dividing by 1. A sort of *"all the other numbers have got their own series so why shouldn't 1 have its own as well?"* Might sound OK if maths was a Soap, but maths isn't a Soap.

The definition of a prime number, however, helps: *a number that can only be divided by itself and by one*. Division by 1 is an accurate description of these triples, yet mathematics has held a strangely ambivalent attitude towards them.

I will look at that attitude in the next chapter. For now, I want to look at Division by 1 a little further. The generator doesn't only give a working reality to the definition of a prime number, but the functioning of this first series governs the entire system.

At this point I need to give some simple terms:

- To distinguish the three numbers in the triple I've called them **Minor Value**, **Median Value**, and **Major value**.
- **Sin MiV**: When I started looking at Pythagorean triples I realised I would need a ready way of tagging each one. The triple is a three-way ratio, so if all three values of the triple are multiplied by the same (whole) number it will be another triple but its ratio will remain the same. I decided to calculate the sine of the minor angle for each one and this is given by: 'Minor Value ÷ Major Value'. I always refer to it as Sin MiV and it's the only tag I've used.
- **Divisor**: all numbers will be divisors. The value is the difference between the Median Value and the Major Value. In Chapter 4 that is going to have to be modified, but it can stand for the time being.
- Every divisor has its own sequence of dividend values, with a specific step from one to the next. I have referred to this step as **Division Interval**.
- For the sake of brevity, I will sometimes abbreviate a specific 'Division by..' number, say Division by 3, to (÷3).

There are other terms and, as before, you can look at 'Bones and Biscuits' to get more detail but if you want to stick with this, then these are all that are needed for the moment.

Every number can be divided by 1 and for the most part it's a pointless thing to do. But in the generator all divisions take place on the square of the dividend value, with the objective of finding where that square, or the set of odd numbers equal to it, lies in the line of differences. That's what was done in the first example at the start of Chapter 1 and it's that 'set of odd numbers' that matters.

It is just as true that any square can be divided by 1 and, done in isolation, it is just as pointless. But the generator doesn't do 'pointless'. The squares of all even numbers cannot exist in the line of differences as single values. The square of an even number must be another even number. But the differences in the x^2 series are the odd numbers. 'Division by 1' does have meaning because it locates those squares which are single values, the odd number squares, and it identifies their triples. It takes every odd number in sequence, and therefore has 2 as its division interval.

This division interval of 2 and the process of forming the triple produces a further, important, effect. The difference between the Minor Value and the Major Value becomes a sequence of: *2 × the sequence of squares*. These are listed, for the early stages, in Table 3.2. The first is 2×1^2. Obviously that is 2 and something that looks distinctly punctilious. But the following value is 8, then 18, 32, 50, so starting off with the double of the first counting number squared, and seeing it that way, doesn't seem quite so pernickety.

When these values arise in the 'Division by..' sequence as divisors they work on a finer division interval than might be expected and produce new ratios of triple. Going back to the terms I gave earlier, amongst the triples which these divisors produce there are Sin MiV's which have not arisen previously.

There are further important characteristics in the generator, in particular in Division by 2, but they can keep for a while. What I have shown here is that dividing by 1 can have real meaning and it does so in The Triples Generator. Many of us have at one time or another recited the definition of a prime number without ever thinking whether it might have a real meaning. Could some thought around the definition have produced a worthwhile result?

Perhaps. The first observation might be, *"We use this expression, 'can be divided by 1', but we never actually do it because it doesn't have a real meaning. Are there any circumstances in which the phrase could have real meaning?"*

From there: *"It would have to do something that identified a single entity but that could only have relevance if other numbers were working to identify entities which were not intrinsically singular but were in the same series."*

"The number series would provide the condition of a sequence of entities that can be divided but that is the factorisation sequence, and that's where we started – the number 1 doesn't have a real role. What if something is done to the number sequence, say, squaring it?"

"The difference of the sequence of squares is the sequence of odd numbers. We know the square of every odd number is an odd number and is therefore in the differences as a single item. The even squares

exist as pairs of differences. What if some squares can occupy more than two differences?"

"Sure, there must be some but how far would it go and what would it achieve?"

"Dunno. Let's go and have a look."

That is the first, and very hypothetical, shift of thinking that might have happened and would have resulted in finding The Triples Generator. There are others to come later. It may be argued that there is an unreasonable level of hindsight in formulating this scenario. But then look at the conversation I have suggested. Would it really have been so difficult to have formed the first question and then worked through that line of questioning and reasoning?

At this point the attack dogs might be heard, *"Mathematics knows quite enough about Pythagorean triples and the x^2 series and doesn't need to know any more."*

"Really?"

In the light of the response I got to saying 'infinite division matrix' such a reaction has to be a possibility. This might be a good time for the dogs to quieten down.

CHAPTER 3

BEYOND DIVISION BY 1

Although division is the main arithmetic operator in the generator, because it is being done on the square of the dividend value it has the effect of multiplying earlier generated ratios of triple.

The first series in which this effect can be seen is Division by 2. This series works on a division interval of 2 and produces a repeating sequence of: *a 2× multiple of the corresponding triple from Division by 1 followed by a new ratio of triple,* ie a Sin MiV that did not arise in (÷1).

It is worth having a look at that other characteristic, the difference between the Minor and Major Values. Where Division by 2 produces a new ratio, the Minor/Major difference is the square of an odd number. Where it multiplies up the ratio first formed in Division by 1, the Minor/Major difference is the square of an even number. Table 3.1 is a listing of the first twenty five triples in the Division by 2 series. Table 2.2 shows how these new ratio divisors are identified in the x^2 series.

Thus, the double of all squares and the odd number squares produce new ratios of triple, when these values become divisors. The depth and character of the generator is established in these first two 'Division by ..' series. The even number squares from the Major/Minor differences in (÷2) become divisors which multiply, by 2, the ratios from the corresponding value in (÷1). So, Division by 16 doubles Division by 8 and (÷36) does the same for (÷18), and so on.

There are a lot of numbers which are not squares or double squares. They all work, in sequence, as divisors in the generator. In general, the odd numbers multiply (÷1) by their own value; the even numbers multiply (÷2) by half their value. Where the ordinary numbers are multiples of a 'new ratios divisor', that's the one they multiply. The start of these relationships can be seen by looking through Tables 3 and 4. However, just as The Triples Generator can place triples in the x^2 series but doesn't have to, so the same characteristic applies to the reader and these tables.

I remarked in the previous chapter that mathematics has had a strangely ambivalent attitude towards the Division by 1 triples, the ones of the form 5,12,13. That ambivalence is characterised by giving this class of triples a name, the only class of Pythagorean triples to be given one and, having done so, to then trivialise them. In the 1960's, when I last had any contact with the Pythagorean triples, they were known as 'the limping triples'. As a young schoolboy I hated the name from the moment I first heard it. I didn't know why. I just did. Later I found out that my childish gut reaction had been spot on.

Seemingly the term dates back many centuries and quite probably to the time of Pythagoras. Incidentally, he did not discover the Theorem of Pythagoras; it was known at least a thousand years earlier to the Babylonian civilisation and quite probably stands, like the wheel, as an unattributable fundamental of human development. Pythagoras and his followers sought to explain the world in numbers and this was their crude and nasty way of 'explaining' a defect of the legs.

Something has happened in the past forty years. It seems that mathematics has decided the word should now be replaced. This being the recent past, the end of the 20th century, one expects a certain degree of political and social awareness. This is the time, at least in the developed world, of a generation that grew up to overthrow many of the ways and attitudes of the past:

"OK, fellow math–men and –women, there seems to be general agreement that we should get rid of this 'limping'. Any suggestions for a new word."

Did anyone, ever, ask why these particular triples had to have a name at all? Were there any suggestions and, if so, what might they have been? Remember, these triples are an odd number and then two consecutive numbers, so we might get: 'the odd triples' or 'odd number triples'. Other possibilities: basic, fundamental, unitary, elemental, primary, elementary, simple, even mathematics' very own and deeply loved word 'trivial' might have got in. But no.

These triples are now known as 'The Primitive Triples'. Like that's an improvement? Have these people never heard of the loaded word? If they have, then what are they doing messing around with this one? If they haven't heard about it then on which planet did they go to school?

Even today mathematicians still refer to these triples as having 'legs', when what they are referring to is the Median Value and the Major Value.

Why then trivialise them? The naming is an important part of the trivialising process but it isn't, in itself, an explanation for trivialising. The reason is amazingly simple: they are easy to find. Mathematics loves 'complex' because that is what mathematicians are good at. And, yes, they are very good at 'complex'.

If you look at a website or book dealing with Pythagorean triples you will be shown how to find one of these triples. You will then be told, either literally or by implication, that they are the trivial cases. You will be led off down the lanes of investigation that the existing formulae and techniques provide to find what mathematicians judge to be the interesting triples and what their techniques have determined to be interesting groups of triples. And what is being done may indeed be enlightening and interesting.

But this process also means that mathematics has turned its back on simplicity. Their own language, 'the trivial cases', powerfully reinforces their attitude. Earlier, when I showed how 91^2 can be found in the line of differences and its triple of 91,588,595 easily identified, I made the assertion that if one triple can exist in the x^2 series in that manner then all will so exist and can be so identified.

That assertion applies just as strongly to identifying these single value, Division by 1, triples. It is neither rocket science nor Holmesian powers of observation, to see these, and the Division by 2 triples using a pair of differences, standing in the x^2 series. Mathematics knows perfectly well that they are there. The shift of thinking is to ask whether that could be the early stages of an infinite series.

CHAPTER 4

THE REVERSE TRIPLES AND NEW RATIOS

Each of the new ratio divisors starts its sequence of triples with one or more in a reverse form. Instead of the triple being in the normal sequence of, Minor Value, Median Value, Major Value, these arise in the sequence of Median Value, Minor Value, Major Value. In the early stages of looking at the generator it presented me with a naming problem. A column headed 'Minor Value' would be starting off with the Median Value of the first triple and, in some cases, a few more to follow.

I solved that problem by heading the columns 'Addition Value 1' (AV1) and 'Addition Value 2' (AV2). The AV1 is always the value whose square is divided. I mentioned in Chapter 2 that this modification would arise.

With the realisation that the reverse triples had to be seen as part of the generator's functioning it became necessary to see the divisor's value as the difference between (AV2) and the Major Value.

A look at these reverse triples shows that they have all arisen in earlier 'Division by..' series. The following are a few examples:

In (÷2):	4;3;5	First arising in (÷1) as:	3;4;5
In (÷8):	12;5;13	First arising in (÷1) as:	5;12;13
In (÷8):	16;12;20	First arising in (÷4) as:	12;16;20
In (÷9):	15;8;17	First arising in (÷2) as:	8;15;17
In (÷9):	21;20;29	First arising in (÷8) as:	20;21;29
In (÷50):	60;11;61	First arising in (÷1) as:	11;60;61

It is obvious to see these as the reverse form of an earlier triple. The divisor that forms the reverse triple has the same value as the difference between Minor and Major Values when the triple was first formed in its normal order.

Every triple is a sequence of numbers. Therefore there will always be a number that is the difference between Minor and Major values. At some point that number will become a divisor. When it does so it will have to divide on the square of the median value. This median value was identified when the new divisor's value was still a number and was the Minor/Major difference in the earlier series.

It may seem strange to see the triple 11, 60, 61 being reversed as the first triple in Division by 50. However, just as dividing 11^2 by 1 had to identify 60^2 in the line of squares, so 50, the difference between 11 and 61, has to divide on 60^2 when it becomes a divisor and identify 11^2 in the line of squares.

(It should be noted here that the processing of the divided value actually identifies the second addition value of the triple, ie the 11 or the 60 in the above cases. How this arises is dealt with in 'Bones and Biscuits'. At this point it is quite enough for the reader to know that there is a line of squares and a line of differences. Each square has its root number sitting directly above it, so there doesn't need to be a problem here.)

Whilst a human view will always be that 60,11,61 and 11,60,61 are the same triple, the generator, having no consciousness, produces what it does because that is what can be produced at that point by that divisor. Thus all triples will arise as reverse triples and when they do so it is because that is the only possible outcome.

Table 5 is a listing of all the triples shown in Tables 3 and 4 and which can then be found as reverse triples elsewhere in these same tables. I decided to look at two somewhat larger diviors: Prime 1009 and its square, 1,018,081 (See Table 3.8). In the ($\div$1009) table I have shown which divisors will reverse each of the triples generated. In the ($\div$1018081) table I have shown where each of its first twenty five triples, all in reverse form with a lot more to come, originated.

The crop of reverse triples has an increasing Sin MiV. Having reached a peak of ratio value, the triples are then formed in normal order and the Sin MiV reduces with each triple generated. In some instances the greatest ratio value is at the last of the reverse triples; in others it is at the first of the normal order triples, and, in ($\div$72), it is balanced across the boundary.

There may be a seriously complex mechanism here. There probably isn't. Reality is that this is what each new ratio divisor does on these lowest of its dividend values.

When the new ratio divisors begin to produce normal order triples, after the reverse triples, they produce new ratios and multiples of previous ratios in a sequence that appears to be unique to each of these divisors. From the relatively small range of divisors covered in my study there does seem to be a characteristic in common for the prime squares. Divisors 9, 25, and 49, each produce a sequence of new ratios which is one less in number than their root value, followed by a triple with a Sin MiV from the (÷1) series. This characteristic may well be unique to prime squares because (÷81), the first non–prime odd number square, has a sequence that is quite different.

But what are all these new ratios for? Do they signify something? I think they probably do. As each one is generated it is defining the minimum size of whole numbers that obeys the Theorem of Pythagoras and has that particular ratio. The ratios interleave the points first laid down in the (÷1) series. Some of the new ratios exceed the ratio of the first triple in that series, 3,4,5 (Sin MiV: 0.6). These can be seen as interleaving between that value and the maximum possible Sin MiV of 0.7071 (sine of 45°).

If there is significance in the new ratios then it must be more than just the ratio. If the ratios alone are interleaved then all it would give is a very finely graded, first half only, sine table. Here, any investigation would switch the priority from the triple to the ratio, with the triple becoming the tag. I have not attempted such an assembly in my study as I doubt there is sufficient data to give a meaningful analysis.

That is, I am sure, true. Both about the value of doing a study along the suggested lines and there being insufficient data in this book. However, on this one I would not be at all sure how to look at the problem. I'm sure about noting the size as each new ratio arises, but I have a strong feeling that the size tag may need to be something more sophisticated than a simple statement of the triple. I'm certain there is something in this area of the work but I'm not your man to find it.

CHAPTER 5

THE GENERATOR AND PRIME NUMBERS

As I worked through the new ratio divisors I started to become aware of prime numbers in the new ratio triples. They arose quite often in the Major Values. There is only a small amount of data. Table 3.8 summarises what is there.

In the (÷1) series the generator uses all prime numbers as the Minor Value of a triple. But it makes no distinction between a prime and an ordinary odd number. No other divisor can use a prime as a Minor Value.

The divisors which are key to isolating primes are those which are prime. The prime square divisors, working on the same interval as their prime root, do the same isolating job as the prime divisors but with much fewer divisors involved.

Once I got into a method to isolate primes, it became clear that the prime square must be seen as a stage in the progress of the prime divisor, now to become a prime identifier. The prime square divisors perform an important role in the generator, but once the method of the generator is switched to isolating primes, the prime square divisors no longer play a distinct role. However, the prime square position for each prime identifier is very important.

The raw mathematics to come in this chapter will be kept as simple as possible. There is a data summary of the test in 'Bones and Biscuits'.

First, a few words about the ancient Greek mathematician, Eratosthenes. His simple and effective method of isolating primes is known as The Sieve of Eratosthenes. Each number, starting with 2, crosses out all the numbers it can factorise. Thus, 2 does the evens, 3 does the odds factorised by 3, then 5, 7, 11 and so on until all the scribes have got RSI. After all this crossing-out the numbers left standing are the primes.

The method uses lowest prime factors only. Each identifier must work from its own square onwards.

My starting point of The Triples Generator was helpful because it has to be seen as a matrix. Eratosthenes showed that all you need to do is write out the number series and then start the crossings-out. Both concept and method work in a straight line.

The matrix concept looks like this:

	Multiplier Index (odd number series)												
	3	5	7	9	11	13	15	17	19	21	23	25	27

	The Prime Identifiers												
	3	5	7	11	13	17	19	23	29	31	37	41	43
3													
5													
7													
9													
11													
13													
15													
17													
19													
21													
23													
25													
27													
29													
31													

The Main Index — odd number series

On a somewhat larger size, that was both concept and method to isolate the primes to 109999, the test limit I set. The identifiers required are set by $\sqrt{109999}$: the primes from 3 to 331, a total of 66. By setting up an odds only Main Index, Prime Identifier 2 had, in effect, been used.

Each identifier is set on its step value (2×Identifier). It is a summation process and marks off each of its odd number multiples up to the stage limit. So far, apart from not doing any crossing out, there is no difference to Eratosthenes. At the end of identifying all the non-primes in the range, the primes are the numbers on the main index which have no entry against them.

The multiplier index is conceptually useful. Factorising of the marked value on The Main Index can be due just as much to the multiplier as the identifier which actually does the marking. Later on the multiplier index is stripped to its bare bones.

It is helpful to keep this index as a separate entity from the main index. It controls the progress of each identifier whether it is working in the summation fashion or it is producing explicit multiplied values.

By running out each identifier separately, their pattern becomes clear. Identifier 3 produces the prime pairs, which are soon disrupted. The first two disruptions are by identifier 5 at 25 and 35, the third is by identifier 7 at 49.

The first 13 odd number primes – the ones from 3 to 43 – were run in the summation fashion.

The $\sqrt[3]{109999}$ is 47.91.... All identifiers lying between the cube root and the square root of the test limit can be run against a 'primes only' multiplier index. By allowing in 47, just below the cube root, I had to include its own square, 2209, in this otherwise 'primes only' multiplier index.

The multiplier method produced just over 5K of numbers and each is unique. I refer to these numbers as 'non–pS'. They can be seen as a distinct class. They are dependent on the notion of a test limit but that can be as far away as one wishes.

In the broad brush analysis to follow I am quoting on an 'all numbers' basis.

The number of primes isolated, correctly I might add, were just under 10.5K (9.5%). The 5K of 'non–pS' numbers was 4.7%. It is useful to put these two percentages together, 14% will do. The remaining 86% is made up of 50% marked off as even numbers, by Prime Identifier 2, and 36% marked off by the next 13 primes.

Throughout the number series the 86% due to 2 and the next 13 primes remains the same. The primes and 'non–pS' numbers will be the remainder. As the number series goes on the quantity of both classes becomes considerable but, remember, the quantity of numbers also becomes considerable.

The method I used produced much fewer markings–off than if all the identifiers were run in the summation fashion.

If this system were to be run to the next stage, the split at ($\sqrt[3]{}$test limit) means many more identifiers would run in the summation fashion. The 86% would increase but the number series will still accommodate the increasing quantity of primes and 'non–pS' numbers in the remaining proportion.

The highest prime isolated was 109987. As an identifier it will factorise, ie mark off, fewer than five odd numbers per million of all numbers. If the last 5000 primes isolated in this test were averaged to find their collective marking rate as identifiers, then all of them together will mark, roughly, 60K odd numbers per million all numbers.

Identifier 3 marks at the rate of 167K odd numbers per million. Small primes have a very high intensity. But what is small?

CHAPTER 6

A SAND BERM BIGGER THAN THE UNIVERSE

I recently read that all primes less than a thousand digits long are small. Elsewhere the website talked about isolating primes. The 17digit numbers were recently completed and slow progress is being made through the 18digit numbers. The bad news: there is insufficient capacity to store all of the primes identified.

The two statements are in stark contrast. Who made, and how, the definition of 'small'? What exactly does '....digits long' mean? Mathematics shows specific large numbers accurately but when talking of large numbers in general the '....digits long' handle is the norm.

Try this out. Take the number 1,000,000, the first 7digit number and just one more than the total quantity of all numbers before the 7digit series. Multiply it by 9. That is the total quantity of 7digit numbers. The '9' multiplier is a handy guide to what digit length really means. Nine times more than <u>all</u> previous numbers.

Where the term '....digits long' is freely thrown about, it does help to give a sense of reality.

"But surely, this characteristic of the numbers must be understood by mathematicians?" Of course it is. But at what level?

Big primes form their factorisation sequence in exactly the same way as small primes, and all the ones in between.

What has changed with the big primes is the space between each of its multiples. Consider a prime arising in the range of 40digit numbers. Its first multiple in the odd numbers, 3×itself, will be either 40 or 41digits long, but still three times bigger than a very big number. Its square will not arise until the 79 or 80digit numbers.

We are used to numbers like 40 and 80. But with the word 'digit' attached, these numbers need to become unfamiliar. The 41digit numbers are a 9 times greater quantity than <u>all</u> of the numbers up to the end of 40digits. And then again for 42digits, and again, and again. Now the distance from 40digits to 80digits is beginning to look real. I am going to risk labouring a point because I am sure that, once one gets into the '....digits long' region of the number series, there is no real understanding of this simple fact: **when the digit length is increased by 1 the sequence started is 9 times greater than all previous numbers.**

Euclid proved that there is an infinity of primes. Gauss showed that the proportion of primes to all numbers had to reduce through the number series but could never be zero. Thus, a slightly reduced proportion of a nine times greater quantity is still going to be a massive increase in the absolute quantity. And with each unit increase, the numbers themselves get bigger as well as the number of numbers – hence the difficulties with storing all the primes in the 18digit series.

Whilst quantity of these numbers is their most significant and most neglected attribute, what each one of these big numbers means is also not easy to understand. Looking again at a 40digit number: imagine it is a distance in miles; imagine that it is possible to make a day return trip from earth to the sun. There are not enough days in a billion billion lifetimes to cover that many miles.

Is it believed that primes up to 1000digits are 'small', whilst also knowing that the non–primes around them are seriously big? And a seriously big quantity of them?

Whatever the case, no matter how deep or stark the disconnect, nobody will ever send a squadron of number searching tanks into this sand berm. Probes have been sent into the berm and a large number of primes have been extracted. Large that is, in relation to the low end numbers we live and work with. As a proportion of the sand berm the quantity found doesn't register. Even as a proportion of the primes in the regions of the berm from which the samples have been extracted, they still barely register.

CHAPTER 7

PRIMARY PURPOSE

The primes are a big deal. There are two main reasons for this: their use as the basis of the public key/private key encryption system; the lure of proving the Riemann Hypothesis.

The first involves money, a great deal of it, passing through the international economy, and now coming down to ordinary people shopping on the internet. When money and the public come together it is not difficult to build up interest. If the encryption system had only ever supported the transactions of banks, governments, and big business then prime searching would have remained where it has been all along – in the world of Number Theory.

Attempting to prove the Riemann Hypothesis has remained in that world. Its profile has been raised but just to understand what the hypothesis states is well beyond a simple word form we can all recite. The encryption system will face many challenges. Finding a sufficient stock of primes to keep it going is not one of them.

Number Theory as a distinct discipline of mathematics has developed over the past two hundred years. The work of Gauss, demonstrating the connection between primes and logarithms, is seen as the key point, in the early nineteenth century. Euler, in the preceding century, produced some valuable techniques. Riemann built on the work of both Gauss and Euler to produce his hypothesis.

This branch of mathematics deals with whole numbers. The dynamic of study is focussed on the primes, seeing them as separate from all other numbers. In looking at other aspects of the numbers, the theorists have mainly sought to inform their study of the primes.

The intention of all this study has been to find rationality, a definable system, by which the location of all primes can be stated. If there is ever to be such a ready reckoner, then the best hope for it lies in the method devised by Riemann.

Some of the record so far, post-Riemann, is:
- 150 years of study.
- Many different methods developed to identify primes.

- Large number of primes identified.
- Different methods developed to test numbers for primeness.
- At least twenty different classes of prime numbers identified.
- A large body of correlations between Riemann's method and the incidence of primes.
- Identification of primes and prime patterns with some naturally occurring phenomena.
- Identification of a large number of prime gaps (sequences where there are no primes) in the big numbers.
- Identification of prime pairs in the big numbers.
- Invocation by a seriously important mathematician at the start of the twentieth century, one David Hilbert, to prove the Riemann Hypothesis.
- The development of a number of primes-based coding systems of which RSA is probably the best known.
- Offerring of a prize of $1,000,000 at the start of the twenty-first century by a seriously wealthy math fan, one Landon T Clay, for proof of the Riemann Hypothesis.
- No proof of said hypothesis.
- Estimates of proof gestation run to 500 years.

Such is the nature of maths proof that one day we don't have it, the next we do. If anyone is getting close even their best friend doesn't know it.

But what would proof of the hypothesis do? Would it really be a ready reckoner for all primes? If so, could it be reversed, provide a ready cracker of primes and bring down the internet and a lot else besides? But Riemann's method is said to require considerable delicacy in handling. As it stands, it doesn't look like becoming, even with proof, either ready reckoner or cracker.

Would a proved hypothesis inform other areas of mathematics, perhaps even the sciences and technology? Without the thing being proved it's impossible to say. However, with many 'unprovens' it can be indicated what kind of effect the proof might have. Looking around the Riemann Hypothesis – well let's say, you're not lost in a forest of signposts. In fact the only really firm signpost is the one that says, "This Way To More Primes". Is the study of prime numbers a closed circuit, a self-perpetuating system?

An exposition on the primes will start with a brief explanation of what they are, involving small numbers. As with the triples, you are rushed off the simple numbers so as to get on to the important stuff. Mersenne primes and GIMPS get in; links with naturally occurring phenomena; then the transaction security stuff, and how this is why we need to prove the Riemann Hypothesis. Finally, *"....and why do we need to know all about these prime numbers? Because they're the building blocks of big numbers!"*

Mankind does not do big numbers, the ones with the '....digits long' handle, very much. When they do arise in the real world they are dealt with. And those who deal with these numbers do not go around asking each other, *"What are the building blocks for this one?"*

Come to that, what of numbers that don't divide into whole numbers? This is not one of the crashing points of civilisations. Primes have never halted, deflected or slowed a construction project, an IT project, or any other kind of project.

Could number theory have given itself a wider horizon? Could it have asked questions of itself? Many of the methods developed to identify primes do so by showing them to be a residue, an unused number. Could this have caused a shift of thinking?

"If primes are a residue, then what kind of system would discard them?"

"Obviously, it would have to use division as its arithmetic operator and the system would have to be infinite."

"If it is going to discard all of the primes then it must use every prime as a divisor and make all numbers available as dividends. It would probably end up having all numbers as divisors."

"We've got the Sieve of Eratosthenes. It does fulfill those criteria but it is a method to find primes. It isn't an other-directed producing system."

"What if we did something to the numbers, the dividends that is, say squaring them, and then divided the squares?"

The presence of primes in the number series has held a deep fascination for man. We bring our desire for order and our standard of what constitutes order. We find that these numbers stubbornly refuse to conform and then they compound their stubbornness by refusing to form an order of unconformity.

The number series is formed by an *"....add one and three, add one and four, add one and...."* process, because that is the simplest method. The complexities become visible because of the human use of the series. Such is the rigour of our usage, we know that the complexities are inherent. However, they are neither formed within the creating process nor are they inherently visible.

The Triples Generator makes the primes invisible. Where they are included as Minor Values, in Division by 1, they are indistinguishable from the other Minor Values in that series. In the remainder of the 'Division by ..' series primes are unused as First Addition Values (AV1) and their unuse is unnoticeable. Even if there is a greater proportion of primes arising in the Major Values, then their presence carries no demand for recognition.

If the evidence points to the residue status of prime numbers, then the implicit belief that studying the residue must ultimately lead to the ordering of the primes says a great deal about faith and determination, even obsession, but the physical world around us gives few, if any, auguries for the success of such an approach.

Once a system such as The Triples Generator has been found, it is very unlikely to give any greater insight to the primes than all of the work so far undertaken by Number Theory.

That may well be the generator's true contribution. It encourages the view of the factorisation sequence as a matrix. This underlines the overwhelming importance of the small numbers in determining what will remain prime.The newly discovered largest prime can only perform one function: to contribute the tiniest possible element in identifying all primes greater than itself. As there is no chance of all previous primes ever becoming known, then even this, the tiniest possible value, is nullified.

Small numbers: small quantity, big factorisation effect.
Big numbers: big quantity, small factorisation effect.

The real change needed in this striving to find order is to realise that we are seeking to impose a human and very restricted sense of order upon these numbers. They clearly have their own order. They are not intractable, they are not random and they are certainly not chaotic.

If one uses the concept of the matrix to isolate primes in the number series, then the identifiers can be seen as a phalanx. Each identifier is working through its own, unique, factorisation sequence. Yet it is their effect as a phalanx that will isolate the next prime. Once isolated, the latest prime has to be attached to the *ad inf.* end of the phalanx. It thereby makes the phalanx different to what it was before. When the numbers become big, then the change is tiny, but changed it has to be. There is systemic impossibility in seeking our idea of order in the primes.

The isolating of primes, starting with 2 isolating 3, then 5, then 7, is infinite. There is no barrier in the numbers, no inherent stop position.

If we know that this universe contains neither enough matter nor enough time for this pursuit, are we not obliged to accept those facts and learn to live with them?

The number series makes no intrinsic demand for the primes to perform division, ie to act as identifiers. Furthermore, there is no inherent need for division in the number series. But The Triples Generator uses division as its principal arithmetic operator and it therefore should be seen as the identifying source of prime numbers. The shift of thinking is to recognise that non-use, by all the divisors greater than 1, is the means of identification.

In time to come, this obsessive pursuit of a few, ever larger primes will be seen in the same light as Middle Ages theologians debating angels on the point of a pin. But at least the theologians were conducting a debate, and not a childish, macho, *"look how big it is now"* game.

I noted earlier that the prime square divisors might have a consistent pattern in their sequence of producing new ratios of triple. It may be that the pattern is unique to the primes, at the remove of their square.

This would be a layering of irony. The Triples Generator is a natural derivative of the x^2 series. Here, in the detail of the system, the primes, at one remove, may be stepping into an order that arises naturally, is unique to themselves, and is a clear ordering. And then be not the least bit of help in predicting where the next prime will arise.

A couple more useful notes for the fridge door:

PRIMES – *"Had we but world enough and time...."*

REMINDER – each number is unique

CHAPTER 8

ON EULER'S CONJECTURE

Euler conjectured, in 1769, that there would be no whole number solutions to a range of equations where the index to which each element is raised is the same as the number of elements.

The equations are of these forms:

$$a^4 + b^4 + c^4 = d^4$$
$$a^5 + b^5 + c^5 + d^5 = e^5$$

In 1966 Lander and Parkin found the following equation, thereby disproving the conjecture, at the 5 level:

$$27^5 + 84^5 + 110^5 + 133^5 = 144^5$$

In 1986 Professor Noam Elkies found the following equation:

$$2,682,440^4 + 15,365,639^4 + 18,796,760^4 = 20,615,673^4$$

thereby disproving the conjecture at the 4 level. Seemingly, Euler had made a special point about the quartic of this form, so Professor Elkies' discovery is seen as particularly significant. He also proved that there are an infinity of these equations.

In 1988 Roger Frye found what he proved to be the smallest possible equation of the 4th power:

$$95,800^4 + 217,519^4 + 414,560^4 = 422,481^4$$

Since these discoveries many more equations have been found, with a web cottage industry developing around them. Having looked at many of the sites I cannot find anything that suggests a systematised 'dividing the quartic' or 'dividing the quintic' approach has been taken. It could be worthwhile having a look at this.

For this purpose I will return to the '$a^4 + b^4 + ...$' form shown earlier. What follows can be applied equally to the quadruple and the quintuple, but I will refer mainly to the quadruple.

I need to lay out some terms of reference. The three values on the left of the equation are the addition values. The one on the right is the major value.

One, and only one, of the addition values can be held in the line of quartics at any one time. The other two addition values will lie in the line of differences. These can only exist there in their enhanced state. So if you were thinking of 95,800 in the line of differences it would be as $95,800^4$ (84229075969600000000). The major value can only be held in the line of quartics.

The following configurations can arise:

- b^4+c^4 in differences; between a^4 and d^4 in quartics.
- c^4+b^4 in differences; between a^4 and d^4 in quartics.
- a^4+b^4 in differences; between c^4 and d^4 in quartics.
- b^4+a^4 in differences; between c^4 and d^4 in quartics.
- a^4+c^4 in differences; between b^4 and d^4 in quartics.
- c^4+a^4 in differences; between b^4 and d^4 in quartics.

The values in the line of differences have to be seen either way round – the differences are progressive. Each pairing shown must have a shared difference. If not, the differences would be a complete, single, quartic value and would therefore produce a triple – now a proven impossibility. Thus, there are a total of up to six values of difference which can be shared by whichever two quartic values lie in the line of differences.

For the quintuple, the arrangement becomes very much more extensive. Three quintic values lie in the line of differences. There are four primary configurations and six possible arrangements giving up to ten shared differences.

I have started to look at the way in which a 'Division by..' process would work. The divisor required would be given by the difference between the major value and the addition value in the line of quartics. For example, if a^4 is in the line of quartics, then the divisor would be (d – a). I don't know how to get from the quotient to the lowest difference value to then check the addition value in the line of quartics. But the first step is to see if the summation of the quartic values lying in the line of differences produces a whole number qoutient when subjected to the appropriate divisor.

They do, in each of the three main configurations for both the quartic equations shown earlier and for each of the four main configurations of the quintic equation. I also checked that even divisors produce even qoutients and odd divisors produce odd quotients. That means it is likely that the summed quartic or quintic values could be distributed along the relevant section of the line of differences. From the point of view of doing the division, the sequence of the two values in the line of differences, doesn't matter. The sequence might matter, and the shared difference value, when determining which two quartics will combine to equal the summation of differences.

Two points need to be made. Firstly, showing that each of these equations might be identified in the x^4 or x^5 series, in the manner I have described, is only likely, not certain. Secondly, being given the equations in the first place was a big help!!!

Please don't ask me where to go from here – I haven't got a clue. With three divisors working through the quadruples and four divisors through the quintuple, configurations of shared differences which might be critical to the whole process, finding the next equation, or the previous one, and how, when, and what determines the shared values.....

I'm handing it over. I hope someone wants it and I hope there are generators in these two series.

CHAPTER 9

LANGUAGE AND MEANING

This book is not an acceptable way to present mathematics to a mathematician. If the work can't be presented in their language, it is not acceptable to them. It is the prevailing view within the discipline that anything purporting to be mathematical but is not presented in mathematical language is either wrong or of such surpassing inconsequence that it might as well be wrong.

From this view it is not hard, nor is it unfair, for the outsider to see mathematics' own view of itself as one of perfection. This perfection? Only mathematicians can speak and write mathematical lingo. Therefore, it is only mathematicians who can think mathematically. In consequence, it is only mathematicians who can develop mathematics. The view of self-perfection is inherent in the self-containment of the discipline. And remember, you are expected to ignore the logic breakdown in that syllogistic-form progression.

The mathematicians' reply: *"Learn our language."* For many people that has proved difficult to the point of impossible. The next reply: *"It is a very difficult subject."* That is true, but only up to a point. The 'impossible' barrier comes very early for some.

The expression, '2x + 1'.

The mathematician tells you that this is a linear expression. He means that if you 'let y = 2x + 1', as they like to put it, and then graph the results of that equation, you will get a straight line. (At 'Bones 3' I have looked at this expression and one or two related things.)

He will then say that this simple expression shows the increase in the size of a square. And that is also true. At this point things get confusing. He will start to lay down the law about distinguishing between that which relates to line – one dimension – and that which relates to area – two dimensions. So this 'linear' expression is used to increase area. He then compounds the matter by getting very testy if you exhibit the slightest confusion between 'linear' and 'area'. Brass bound nerve. He started it. He's the one who's confused.

In reality, he never knew he was confused. A very long time ago he accepted the sloppy language and the sloppy thinking that goes with it because this is all down at the low end of mathematical work, pretty trivial really, and in any case, if he got through it then others can and if they can't, they'd be no good at mathematics anyway.

Mathematics does not bother to clear up its own confusion. It does not recognise the inconsistency let alone acknowledge its confusion. Rather, it represents this and other similar inconsistencies as, *"Early experiences of the truly difficult subject that lies before the pupil."*

So, the first step to a mathematics career is the ability to accept the imprecise use of language and definition in some of the basics of the subject. It must be unquestioning acceptance because if the candidate mathematician seriously questioned the discipline's terms at this basic level, then he would be showing himself to be concerned with trivia and could not, therefore, become a good mathematician.

Differentials and differences.

The differential concept chops up an expression into vanishingly small bits, but they are still bits and not zeroes, so that change can be identified and its effect mathematically quantified. Because the chopping up is to a very fine degree some of the expression can be dropped off, eg, the '...+ 1' part of the $2x + 1$ mentioned earlier.

And so the differential expression, nx^{n-1}, is an approximation. Nothing wrong with that – used in the right circumstances, it works very well. Then there is '$(x + 1)^n - x^n$', the other means of finding difference. This is the one from which the nx^{n-1} was derived. If nx^{n-1} is applied to the x^2 series it gives the differences as the sequence of even numbers; when '$(x + 1)^n - x^n$' is applied to the x^2 series it gives the differences as the sequence of odd numbers. Finding which is correct in this very simple case is easy – virtually anyone can do the numbers and find out for themselves.

If the need is to understand difference then the prerequisite is a ready appreciation that the units which make up the difference are of the same order and nature as the units which make up the entities from which the difference was derived.

There are seriously competent maths people who have never understood nor even thought about the distinction between these two tools. If the need is to define change, and then rate of change, the requirement is for a very fine series of differences and they were given a new name, 'differential'. The word is not an arcane construction. It was used to distinguish the assembling of very fine and approximate measures of difference from those which necessarily had to remain substantial and accurate – the differences. Somewhere along the way mathematics has forgotten the distinction.

Multiplication and Addition

The Triples Generator uses the idea of summation in the x^2 series – adding up a set of differences that equal a square and bridge the summation distance from one square to another. It was where I started, wanting to see if there was something in how the series adds up. The generator only uses the first line of difference. If there are similar systems in x^4 and x^5 then, probably, they also will use only the first line of difference.

However, there is an architecture of numbers beneath each of the x^n series. There are lines of difference equal in number to the value of n. Are they merely a prosaic statement of fact?

There is a lattice pattern of addition in all the series, with the values in at least one of the lines of difference being used twice. Again, if you look at Bones 3, this is one of the related things I have looked at.

On the other hand, perhaps you should listen to your friends and get out more.

Anyway, tables 2.1 to 2.4 show the x^2 series in its early stages but they do not include the second line of difference, the recurring value of 2, which as I say, is not needed for the generator. One of the characteristics of arithmetic can be seen on any one of those tables. The balance point between the power of addition and the power of multiplication, given by $2+2 = 2\times2$, appears to shift to 4^2 when the bracketing differences of 16 equal 16. Before that point bracketing differences are greater than the square, after it they are less. This precise progression does not appear to happen in any other x^n series, but then all other x^n series have more than a bare two lines of difference – perhaps that is significant.

If each of the x^n series is just as much a summation as a multiplication series, then the summation can be conducted in two fashions. As I've already said, the first line of difference can be used on its own; alternatively all of the lines of difference can be used. As with my observations on $2x + 1$, only more so, the differences are composed of the units of the series.

Take the x^3 series. The expression $3x^2 + 3x + 1$ defines the first line of difference, so mathematicians say that must be about area. The next one is $6x + 3$, an expression from the '$2x + 1$' family, so that's linear. After that comes the final line, a recurring sequence of 6. On this one it's difficult to get mathematics to be explicit. By their reasoning, they've come down from cube through square and line so I guess the last one is a point. As I say, they are a bit reticent on this one. It's a pity they don't take the reticence all the way through to some fundamental thinking.

A cube is not formed by adding on a sequence of differences that are points, then lines, then squares to the previous cube; it is formed by adding cube units to the cube. The algebra of the declension of differences for each x^n is the servant and, used correctly, the explicator of the numbers. The algebraic declension of differences is scalar in its progression but it is not there so that each line of real, accurate, numbers can be mangled into something it is not, just to suit mathematicians' very restricted preconceptions.

In looking at summation throughout the x^n series, there does seem to be a problem in getting each one started by summation alone. However, the lattice pattern continues for all values of n. Once started, the pattern works perfectly and it might provide insights. Certainly, it would not be tenable for mathematics to regard this architecture of numbers as merely the resort of some poor sap who has not yet mastered multiplication.

Outsiders and Proof

"Surely mathematics must require submissions to be proven?" Why so?

Consider Joe Bloggs. He has had an interest in an area of maths. He forms an idea, a connection that has not, apparently, been seen before. He takes his idea to Mr M and outlines it. Mr M is, quite naturally, sceptical. (Both Joe and I have no problem with scepticism; it's human and it's healthy). In the end, Mr M has to say that he really can't see the connection Joe is trying to make. And then Mr M has to fulfill his professional obligations. He tells Joe that unless he, Joe, can make a proof the idea really cannot be given any credence.

Elsewhere in the world of mathematics, a professor with a very thick wad of credibility tokens goes public. The tokens are the hard-won qualifications and publications of a professional life. Mathematics hands out very few, if any, gongs.

The professor has seen a connection. The paper is published. It demonstrates the connection. It makes clear why proof is not possible.

Mathematics says that an important advance has been made. The connection shows considerable insight. There is a sufficient likelihood of validity that, wait for it, The Bigouad Conjecture has arrived.

So the person who is not able to make a proof, is required to make a punctiliously perfect one. The mathematician who might be able to make a proof is not required to do so, and then receives buckets of praise. Exclusionist culture, or what?

What if Joe was a bird watcher? He identifies what might be a new species. Pictures, notes. Local ornithologist. Scepticism? Sure, and a-plenty. But it stays a live question and, years later and a lot of work by other people, yes, it is a previously unidentified species. Ornithology has no problem crediting Joe, perhaps including 'bloggsii' in the latin name and, maybe, the bird known as 'Bloggs's Warbler'.

(It might be thought that, assuming the generator is a discovery, I'm making a self-serving case here. I'm not. I like the name 'The Triples Generator' – it says on the tin what's in the tin. Besides, I don't much like my surname and using my Christian name would have me reaching for the sick bag, and then I wouldn't like any of my names.)

And what of Mr M? Without realising it, Joe's idea has been stewing in his mind. Ideas are like that. Suddenly he sees it for what it really is and, very importantly, how he, Mr M, can mathematically express it. Now Mr M has a real problem, an ethical and a political one. He knows a paper will result. But how to start it? *"This paper seeks to....Inspiration for this.... Joe Bloggs, a refuse collector in.... ....his spare time reading and thinking about.....*" An ethical attribution of source, but in mathematics – a career suicide note.

Whether he likes it or not, Mr M will have to do what would be the natural response in any other discipline – help the idea come into the world as a useable concept. He has already, unwittingly, evaluated the idea. Now he will have to act as its translator.

Mathematics and Trivia

Mathematicians hold a deep love for the word 'trivial'. For them, it has many meanings.

A mathematician may be explaining a complex concept, taking you up a logic mountain. You're on your rusty old logic bike and he's on this very flash logar–bike. (The logic processes of mathematicians have something of the character of a logarithmic scale – they seem to step effortlessly over swathes of logic steps where the rest of us have to plonk our leaden feet down on every single one.)

Well, today you're doing a reasonably good job of keeping up. He reaches a point for a breather. Here's a function. You can feed in some nice comforting numbers and get something back, and examples are a big help. Nonetheless, it had been hard work getting here and, yes, you can understand where you are but....

And then he says it. *"These, of course, are the trivial cases."*

After a while, perhaps on later reflection, it may become clear why those numbers in that function are 'trivial'. The journey there hadn't been trivial, though. Even the mathematician would not have said that.

So, why 'trivial'? The word's purposes spread like ripples on a pond:
- It is a put-down on the uninitiated (he means it that way, you take it that way, and no one admits to anything.)
- It is both a handshake and a demarcation marker when the mathematician is receiving a visitor from another specialism within mathematics, or from a maths-related discipline. Those from such disciplines are in a special position, ie they might know too much maths, and then the bare-faced but always unadmitted 'put-down' function can be employed.

- It is a badge for wannabes. Something they can pick up and show off to their unmathematical friends. When in the company of maths people, they make sure it is pinned behind the lapel and only discreetly flashed at just the right moment. Sycophancy is welcomed by mathematicians – that was how they got through all those basic level confusions and inconsistencies – but it must not be blatant.

- 'Trivial' sets no-go areas of study. These are deemed to be of such surpassing unimportance that anyone venturing into these areas will be either strongly discouraged out of their intention or strongly encouraged out of mathematics. It is worth noting that most of these no-go areas were set way back in history – the Division by 1 triples are a good example.

- A sub-set of the no-go area are those apparently simple phenomena whose existence mathematicians find irritating – tagging them with 'trivial' makes them go away: the differences lying beneath each x^n series, for example.

All professions and disciplines have their ways of demarcating territory, establishing and maintaining hierarchy. They all have far more snobbery and sycophancy within their ranks than any of their members would ever admit. The long journey of human development has shown that the good qualities and the bad have to travel together. They are in all of us and they are in each of us.

But, and for Mathematics this is a very big 'but', in all those ways of use, the word 'trivial' is being applied to some simple truth. Sometimes it is slapped, like a piece of cheap wallpaper, over the heart of simplicity, and that heart is both sublime and complex.

Again, mathematicians' culture shows a deep disconnect from their disciplinary reality – the greatest proofs are the simplest ones. In their discipline, this is what they admire.

CHAPTER 10

INHERITANCE AND CULTURE

Both the discipline and the culture of mathematics is inherited. In most areas of life the legacy of history has rarely been better than a mixed blessing. For many of us as individuals and as members of a wider society, the history we are forced to carry can be a very heavy and restricting burden.

The history that has created the discipline of mathematics is unique. It is right. Everything ever created in the discipline that is right must survive. If it weren't right it couldn't survive. There is only one caveat on the survival of the right – simple physical loss of documents. The Alexandrian Library burning, over-zealous housekeepers (it seems both Riemann's and Fermat's legacies suffered from this blight), and countless earthquakes, floods, fires and wars have had their effect on mathematics' inheritance.

Learning to live with a history that is right brings special problems and special responsibilities. There is no outside reference register, a means to check by analogy that what you are doing and creating is going in the right direction.

Cultural Imperative
All human institutions and organisations create a culture. Most of these cultures are founded on the core purpose of the organisation but in ordinary life they have to cross-reference to other organisations. For example, medicine has its own special set of ethics and a culture that underpins this but it has to cross-reference to the law and indeed to a quite large range of organisations within society.

Mathematics is no exception to the culture creating imperative. But, unlike most other organisations and institutions, there is no imperative that forces mathematics to cross-reference to other disciplines. Where it does relate to others, mathematics sees itself as handing down (it's definitely 'down', not 'across') the methodology.

So, having a disciplinary history that only allows survival of what is right, and having no need to cross-reference to anything or anyone, it is not difficult for them to build up the self-view as 'perfect'. All of those aspects of basics in the subject where the mathematician is hazy, if not downright confused, have to be cultural. This discipline is, so it would have us believe, the ultimate human thinking machine. How much can we rely on the machine when it cannot consistently state the units needed to increase the size of a square or a cube?

We have to ask how such a peculiar culture arose and how it is so powerfully maintained?

Ancient History

The development of civilisation has always relied on the developing learning of a few individuals. When the application of their learning was successful, they would be lauded and protected by the boss man. When things went belly-up, these individuals needed to scarper – and fast. And so protective organisations developed – craft guilds, freemasonry, the church, though its social and historical role is vastly more complex than just the protection of learning and skill.

Even here mathematics' history is unique. Those who have contributed have survived in name. If one goes back to the earliest named mathematicians, almost all other names to survive are merely the despotic rulers of those times. It would be hard to name a musician before Hildegaard of Bingen yet music has been a part of man's history, probably from our earliest times. Similarly, naming a painter before Giotto is difficult, yet painting also goes a very long way back. Sculpture and architecture can claim some exceptions with names such as Pheidias and Iktinos but these represent the tiniest fraction of the vast range of production in ancient times. Important people such as metal smelters and forgers, stonemasons, wheelwrights and wainwrights, armourers, and even highly prized goldsmiths, do not get to the starting blocks in the ancient naming stakes.

One might think that mathematicians have been numerous in history. In fact mathematicians have been sufficiently rare that there has never been, in all history, any chance of them forming a protective guild, thereby allowing us, and the mathematicians themselves, to see a clear historical dynamic to the development of culture in mathematics. Sure, there have been groups of mathematicians, schools and so on, but nothing that shows a continuing, unifying and protecting culture for mathematicians. By contrast, the City of London Guilds, or Freemasonry did such a good job of protecting their glovers, hatters, masons, whatever, setting professional standards and so on, that they have survived, now with nary a milliner or mason in sight.

Without a formal organisation, or even an informal one, to look at, we have to look to the lives of the individual mathematicians who have come down to us to find some clues to culture. For people with talent the ancient civilisations provided the opportunity for that talent to flower. The mathematics from then was the basis for much development in the ancient world, and it is still important, even invaluable, today.

However, these civilisations could just as easily be violent, repressive, and arbitrary. Ancient times were dangerous for any individual. Luck, and staying on the right side of those who had power were key to survival. The lives of mathematicians in ancient times, though in some cases quite well documented, are so removed from our own times that their cultural legacy has to be seen as distant. It does not readily inform any attempt to understand culture within mathematics in modern times.

Recent Influences

As far as the individual of talent was concerned, did their security improve over the millenia that followed? Not much; sometimes a lot worse. But the lives of recent mathematicans, like all the rest of our history, are a lot closer to our own times and might reveal some of the cultural influences.

But what is recent? What other discipline would look to starting its understanding of recent influences in the seventeenth and eighteenth centuries?

Would medicine defer to Harvey or, later, Lister? Astronomers revere Galileo, but his findings are history – his story now belongs to those who fight repression. What painter would dare to reference the love and anger of Caravaggio, or the sublime delicacy of Titian? Which sculptor would seek to respectfully restate Canova or Michaelangelo? An architect might love Inigo Jones, but would Canary Wharf or The Gherkin be rebuilt on his model? Which engineer would base a new project on Trevithick or Brunel? And will anyone, ever, sculpt with the heedless humanity of Rodin? And that's taken us to the late nineteenth century.

All of these are in the past. Respected, even revered and loved but no longer a basis for new work.

Newton and Leibniz, Fermat and Euler are alive and well, working in the mathematics of the twenty–first century.

Yet Newton wrote in Latin. At the time few could read it; even fewer could understand what the great man said in his Latin. Fermat was a secretive man in his mathematical work; he communicated, but to set puzzles or to tantalise his correspondents. Euler provided great insight and an output that is way beyond prodigious, but on his first appointment in Russia he had to keep his head down for many years. Gauss lived an almost reclusive life in the astronomy institute at Gottingen. When he published he did so having 'removed the scaffolding', as he put it. Riemann essayed briefly in his youth to a more public life and then fell into something very near to Gauss's reclusion in Gottingen.

Even talented people such as these did not make lifestyle choices. Who can say what freedom – so very little – they had to make any choice about how they lived?

Throughout these times social hierarchy was the dominant and domineering characteristic for all who climbed the social ladder. On the one hand, there was enough openness within the society to allow progress for those with talent. On the other hand, there was a harsh and unforgiving structure of social class. Those who depended entirely on their talent could have a hard time of it.

The social context of these relatively recent times does have an influence on mathematics' culture. However, the society of those times cannot on its own provide an adequate analysis of culture in mathematics. There are two other characteristics which are largely peculiar to mathematics and which have deeply influenced its culture.

The Ordering Impulse
The discipline seeks to find definable order. This is a good thing and is, in effect, a restatement of the core purpose of mathematics. The discipline loves its own idea of order. This is a bad thing and the first of the two characteristics. The second is disciplinary tunnel vision. Mathematicians have been known to admit to this. Of course, when they

hear it stated back to them, those who have admitted it in the past will want to deny it. But that's just human beings for you. No one ever truly believes their self-criticism so, when they hear it coming back, it still hurts.

The love of order produces the first and most important effect in establishing a hierarchy of value in mathematics. At the top of the hierarchy is love of the ultimate in 'complex', abstraction. The more a maths specialism can move to the abstract, the more important it is. Those specialisms which have a high level of abstraction are grouped under Pure Mathematics.

Next is Applied Mathematics. In mathematics' hierarchy, let there be no doubt, those who pursue Applied Mathematics are a lesser form of mathematical life. However, mathematicians know that they are no exception to the harsh and ancient law, *"You have what you hold."*

So, Applied Mathematics is held within maths as a whole even though it is primarily concerned with the mathematics of the physical world and human activity. Keeping it within mathematics serves a valuable purpose. Mathematics, Corpus Mathematica, retains ownership of all mathematics.

If Applied Mathematics were allowed to drift off into the world where mathematics is extensively used and thereby diffuse through a very wide range of human activity, different kinds of maths might well emerge that do not belong to Corpus Mathematica. That in itself would be bad enough, but for members of the corpus to find themselves learning some mathematics from a lesser form of mathematical life would be a horror far beyond hell.

Below Applied Mathematics are all those disciplines and activities which make an extensive use of mathematics. Here again, mathematicians apply their impulse to order their cultural surroundings. Someone dealing with atomic physics will be seen as 'higher' than a mechanical engineer. However, there are far too many of these disciplines for a third party such as myself to even begin a Corpus Mathematica Ranking. The corpus not only holds such a ranking but would be nonplussed at the suggestion that these value-judgements are non-mathematical and unnecessary.

In mentioning atomic physics, I have touched on the one maths related discipline that can be held in some regard by the governors of Corpus Mathematica, the Pure Mathematicians. This branch of physics has approached abstraction and as a result a tone of something near to fondness from the mathematicians can occaisionally be detected.

Determination or Obsession

The development of tunnel vision in some areas of mathematics provides an insight to understanding the depth of mathematics' culture. The ordering impulse is powerful but it does not make it inevitable that obsession, tunnel vision, will develop.

Here the study of prime numbers provides what is probably the clearest example. It has been going on for a long time, with very little substantial progress towards the core objective, *tunnel vision(i)*; it has been devoted to studying the primes in virtual isolation from all other numbers, *tunnel vision(ii)*; having got hold of Riemann's method using imaginary numbers ($\sqrt{-1}$), that has become the focus of study, *tunnel vision(iii)*.

The use of imaginary numbers in the pursuit of primes gives a very important character to Number Theory: the specialism gains ranking points because imaginary numbers are seen as having a high level of abstraction.

In fact imaginary numbers, for all their apparent abstraction, serve some very valuable purposes. They arise in the mathematics of electronic and electrical engineering, and in aircraft design.

This mathematics came to you from
AMDA

Applied Mathematics Distribution Agency
MATHEMATICS FOR <u>ALL</u> YOUR NEEDS

AMDA is a licensed agency of the Pure Mathematics Foundation

Never mind that Number Theory must stick with the imaginary numbers. To give up the overwhelming importance attached to the link between the primes and imaginary numbers would be to risk seriously diluting the level of abstraction in Number Theory, and thereby reduce its status within Pure Mathematics.

There is immense difficulty in bringing about a reassessment of any specialism where a serious level of tunnel vision has developed. The specialisms within Pure Mathematics have arisen from the identification of a mathematical area which can be, or has to be, seen as largely abstract. The creation of the specialism is simply common sense. The ordering impulse then invests it with a ranking importance – the higher the ranking the harder it would be to bring about a reassessment if tunnel vision develops within the specialism.

Those inside the specialism see obsession as determination. Beyond that, like us all, they look after their own interests. Others around that specialism may well be able to see what is happening but are almost powerless to act. If the specialism has received a high ranking, then those who might be able to advise and guide on the problem of obsession are effectively barred from saying much, if anything at all.

Ranking by Method

The mathematicians' love of order does not only apply to establishing a hierarchy of specialisms within mathematics. It is clearly applied to the functioning of the mathematics itself: abstraction is better than concrete application, and thence coming right down to simple evaluations of multiplication and division being better than adding and subtracting.

It is clear that the abstraction involved in the pursuit of primes is far from being 'better' than, say, the development of the mobile phone network or a safe and efficient plane, all of which make use of the same relatively abstract entity, the imaginary number. If asked to make a value judgement between the options here then most mathematicians would clearly opt for the practical advantages of phones and planes.

However, it is equally clear that for many within mathematics the discipline's ranking impulse overrides the connection to the real world. When in their world they believe exactly the opposite. Their belief is not founded on an anti-utilitarian analysis, let alone such a philosophy, but rather it is another serious disconnect from reality. The absorption they have made of mathematics' culture is total. It has been a simple imbibing of the notion: abstraction is better than application – the Pure is better than the Applied.

And the reality throughout their world? This is one of the most electronically engaged disciplines. Furthermore, that engagement was in the origins of the electronic revolution. The application of some of the most significant and abstract mathematical thought led to the concrete reality of the electronic age.

Abstraction works in two directions. There is the abstract concept which is then developed into a practical application. Then there is the observation of real relationships which can be abstracted to give the general case, allowing application across a range of situations. These directions of working are well-understood, of equal importance, and are what mathematicians are there to do.

Mathematics is impelled to make a cultural definition of rank and disguise it as a mathematical ordering of importance. To be fair, it is not a conscious 'disguising' but an implicit belief that culture and discipline are one and the same. Thus, the discipline is constantly in danger of throwing the baby out with the bath water. One might come to the view that the work of mathematics is so important to human development that it is unwise to leave it to mathematicians alone.

If you were to ask the mathematician whether multiplication is superior to addition, the answer, the prima facie answer that is, would be that they are different tools for different situations. However, if you quietly open the back door of his mind and peek in, you will see the ranking clearly written on the wall. How else would one explain that they have, seemingly, never found The Triples Generator?

This method ranking plays an important part in the establishing of hierarchy within the mathematicians' minds. It is another feedback in the culture forming imperative, allowing ranking by the predominant mathematical method of the specialism. One which uses sophisticated algebra would be more important than one which uses a lot of number.

"You'd better believe it. The rankings stand. OK?", muttered an attack dog.

The method ranking impulse is given a very clear evocation in the use of 'trivial'. The word is widespread and has become an informal mathematical method. Here, however, there are still distinctions. When applying mathematics for those with serious need, its use is far less likely. In building a bridge or designing a plane there will be some high level mathematics involved. Neither the engineers nor the applied mathematicians are interested in observing that there are 'trivial cases'. There is a job to be done.

In the realms of pure mathematics 'the trivial cases' are quite common. In this area it has, mostly, the handshake and demarcation function. Some popularisers of mathematics use 'trivial' quite freely, offering the badge for wannabes. It seems that a specialism acquires higher ranking the more things it can find which can then be shown to have 'trivial cases'.

It is a strange culture that wants to trivialise its simple truths. What other profession or discipline would treat them with such contempt?

I can find nothing that says where, when, and how this 'trivial' came into mathematics. That is the problem if you don't do a thorough spring cleaning from time to time – the rubbish creeps in and then sticks; eventually you think it's part of the furniture. Perhaps 'trivial' has become the look of choice in mathematics, applied as widely as possible to give an 'en suite' finish.

Mathematics and the Rest of Us

All human beings understand power in relation to themselves and make decisions and pursue goals accordingly. Those human organisations where the existence of political and cultural power is denied are the ones most in need of a political and cultural analysis.

It is strange that there doesn't appear to have been any attempt to understand that there is a culture in mathematics. Perhaps it is all to do with that history of 'the right must survive, the wrong cannot'. It certainly gives a very big inviolability shield to the discipline.

Equally, I find it strange that The Triples Generator hasn't, apparently, been found by mathematicians. But everything points to the fact that it hasn't.

However, it is the history to which I must return. It is in the social context of recent centuries that the final layer of mathematics hierarchy has been placed. And that final layer is the rest of us.

The mathematical inheritance from recent centuries has been total, applying to both the discipline and the culture. The context of a society where the ordering and layering of rank was its definition has chimed with mathematics to form one of the most precise harmonies in cultural history.

The core purpose in mathematics – seeking definable order – has readily translated to a deep love of ordering within the corpus. The mathematics inherited from recent centuries has been profoundly important, both for the discipline and development of the industrial age. Having been produced in a highly stratified society and then going into a discipline that has an inherent desire to do exactly that – stratify and order – it should have been expected that the mathematics and the social context would be inherited as one.

Those stratified societies saw the majority of the population as a mass. The development of industrial society produced many revolutions of which compulsory education was one of the most important. Before this most mathematicians had emerged from the classes above the lumpen mass of people. This was true for all those highly talented people who emerged in European countries over the last three or four centuries. Biographies of the vast range of these talents will often start with phrases such as 'father, a respected craftsman' or 'a family of court musicians' or a lawyer, doctor, whatever. Many others emerged from the higher ranks of those societies.

It is rare to find that one of those talents emerged from the family of a labourer. It did happen and these talents were usually found as a result of work by a local priest or doctor or teacher who took a real interest in the people of his area. But it was not common and one can only surmise how much human talent over the centuries has been lost by the accident of birth.

The motivation behind compulsory education was not some high-minded belief in helping people to overcome ignorance, though it has often been represented that way. Even in the early days of industrialisation, the need for a level of education in the workforce was obvious. It is equally obvious, indeed a necessity, that the education would include the elements of mathematics. It was here that mathematics' culture received a very heavy reinforcement.

Education was seen as anything but, and had been for centuries. The word is from the latin and in essence means 'to draw out' – the same root as 'ductile'. For much of its history, formal education has seen its function as a 'pouring in'. This suited the purposes of the rulers of society and the needs of industry, commerce and public administration. It also fitted well with mathematics' ordering impulse.

What is the purpose of mathematics education? Most would think it is to teach people how to handle numbers. But man has been handling numbers since his emergence as an intelligent, communicating species. There is a small but important fossil record to show that early man used numbers and some of the fossils show an apparent awareness of prime numbers, which would indicate an understanding of division. Much of the finds in early civilisations, Babylonian, Egyptian, Chinese and those of Central and South America show that number was used. More importantly, the archaeological evidence points to widespread use within the population.

So the next point might be: *a highly developed society requires a higher level of mathematical ability in its population than the simple handling of small numbers*. That may be very true but it is not what mathematics education delivers. It would seem that, at best, the general population's number ability is no better than it was a hundred years ago. Any idea of a widespread grasp of more advanced mathematical work is a pipe dream.

If the current level of number ability is all that is needed, then a general education that seeks to develop what is our innate ability would be sufficient. The reality of what is handed down in mathematics education is rather different to that.

The true purpose of schools mathematics is to find mathematicians. It does not have any other function.

If, as a result of the searching method, a reasonable number of people emerge with sufficient skills to populate the maths related areas of work, then that is a bonus. Given our innate number ability, any mathematics teaching method is likely to adequately supply the foundation for the bonus which could then be made more specific in higher education. However, reality is that for many of the developed countries this bonus payment has been declining for some years.

Corpus Mathematica dictates the method by which future mathematicians will be found. Along with its history of that which is right and a profoundly exclusionist culture, the possibility of any serious input from elsewhere towards the education process is virtually nil. This combination of exclusionism and 'history of the right' has produced the inviolability shield and made the finding of cloned generations of mathematicians inevitable – and the rest of us can get on with it.

The existing mathematics education reflects the cultural inheritance, now translated in the opportunity unwittingly offered by a developed society to Corpus Mathematica. The corpus has a highly refined sense of ordering above the lumpen mass of people but the view of that mass is exclusively as the source of mathematical talent. Compulsory education provides the means to find those talents.

The cultural view that Corpus Mathematica makes when it looks outwards at society as a whole is surprisingly, alarmingly, close to the view that an eighteenth century grandee would have held looking out on his society.

Corpus Mathematica does not see itself as having the obligation to provide society with society's needs for mathematical ability in the population. Those needs are always seen by the corpus as so far below the level at which it operates that it can have neither the desire nor the obligation to define society's needs, let alone develop an education system designed to fulfill those needs.

Again, in fairness to Corpus Mathematica, it has to behave in the same way as any other institution of influence in society. It must see its first duty as its own continued existence. The fact that the culture produces effects that many outside the corpus find frustrating and repellent is not going to concern the corpus itself. It has its unique, right, history that has given it an inviolability shield.

Unless society is prepared to make the analysis of what it needs and the analysis of how Corpus Mathematica operates to exclusively fulfill its own primary needs, then there will be no change.

These two analyses would clearly show that, far from moving in roughly the same direction, the needs of society from schools mathematics are so different to those of Corpus Mathematica that the two are likely to be in conflict. At least, if taxed on the matter that is how Corpus Mathematica would see the issue, largely because of its authoritarian and ordering cultural history. The problem for society would be to persuade the corpus that it doesn't need to see itself that way.

If the system of mathematics education is to change – and it will take a massive and prolonged political heave for this to happen – then those with mathematical ability will still emerge. There would be two further bonuses: people in general might see mathematics in a new light; amongst those who were previously put off there may well emerge a truly different kind of mathematical talent.

Popularising Mathematics

There is a contrast in the general population: mention of mathematics will bring out, literally, a groan of dismay from many people; publish a numbers puzzle in the paper and many people will have a go, even some of the dismayed groaners.

This reflects that innate number ability in most of us, and that we actually like having the ability. Beyond this there has developed a growing market for serious books about mathematics. Another reflection of widespread interest that contrasts with the dislike that many people will express for their experience of school mathematics.

Amongst these sorts of books are some which perpetrate the most outrageous hubris on their readers. These authors seek to publicise and popularise their areas of mathematical expertise. They then tell the reader what an awful time mathematicians have with so much rubbish being submitted to them from the general public. Is it too cynical to ask whether it is our interest or just our money that they want?

If mathematicians are being burdened by vast quantities of rubbish from the public, then the following observations come to mind:

- Is it really rubbish or is the mathematician bringing too narrow a perspective?
- If the quantity is large, it indicates that there is genuine interest and thought about your subject out here, even if a lot of it is mistaken.

- If it is all rubbish, then that is prima facie evidence that mathematics is taught badly, and Corpus Mathematica carries the reponsibility for this. The governors of the corpus do not agree that they are responsible? - Read on.
- Don't whinge.

Last Disconnect and Full Circle

It may look like a reasonable analysis to say that Corpus Mathematica is not responsible for mathematics education, at least at the schools level. However, in making such an analysis the cards are being dealt, yet again, entirely in favour of Corpus Mathematica. The analysis has a superficial basis in political reality but to carry it through to the conclusion that the problem of mathematics education is no concern of Corpus Mathematica is to give to the governors of the corpus everything that they have maintained over recent centuries.

This would be a case of the governors having their cake with them for all time and enjoying the eating of it for all time. The 'history of the right' and the ordering impulse fed by an enthusiastically inherited authoritarianism, have been used to create the most impregnable inviolability shield.

The deep reality of the discipline is that it is unquestionably the most 'top down' discipline in the whole of academia. Furthermore, the governors of Corpus Mathematica are required to make no effort whatsoever to enforce or sustain their 'top down' authority.

Those who are below yet within the hierarchy of mathematics look upwards for their direction and for their own relatively small elements of disciplinary, and therefore cultural, authority. They deeply and sincerely believe in the oneness of culture and discipline.

This is a dream for all within mathematics. Beyond there, it is the ultimate dream for all those political operators who harbour the instinct of aurhoritarianism. Dictators and totalitarians, chairmen (they are almost exclusively men) and presidents of organisations inured in hierarchy would pay anything, worship anything, to achieve this ultimate power.

It is, and could only be, in the discipline of mathematics that such power could be a centuries-long sustained reality. The disciplinary history of the right, the illusion of the unity of culture and discipline, and the conscious promotion of the belief that the rightnesses are complete until expanded by Corpus Mathematica, has allowed this power to continue unchallenged for centuries. It has produced a seemingly inviolable, self-regenerating and immutable system of disciplinary, cultural and political power.

To some people, it may all seem to be on a small stage and therefore of little importance. That would be a serious misreading. There is no other area of purely intellectual activity where so few people hold the talent and ability to seriously affect humankind's future.

So, I am forced back to where this book started. Mathematics does not allow any input from outside the discipline. It cannot be approached at any level within its tightly regimented hierarchy with any chance that an input would be heard with a rational, let alone reasonable, ear. The Triples Generator is the input that I wish to make and I am no longer much concerned whether Corpus Mathematica wants it or not.

I am very sure they don't have it, but that is still a long way from certainty. If they do have it, they have kept it very well hidden. Mathematics is not noted for putting any of its lights under the bushel. Those lights which are relatively simple, such as this generator, would be welcomed should they arise from within the corpus, if only to broadcast a newly discovered infinity as yet another 'trivial case'.

Corpus Mathematica may continue to believe that it has a near-sanctified command of all that is right in mathematics. In real life such complacency is rarely sustained. The fact that mathematics has sustained its complacency for centuries does not mean that this has to continue. All things change.

Changing Times

Recent evidence suggests that Peter Shaffer's Amadeus may not be historically accurate. But what the play portrayed, a great talent being dragged down by a lesser, has been a very real fear for those with ability until recent times. This was what prompted Gauss to remove his scaffolding, and he strongly advised the young Riemann to do the same. There was no widespread and reliable system of peer review, nor had there ever been in all of mathematics' history. Professional defensiveness was not only an inevitable response but, in its time, a necessary and wise one.

Times have changed. There are more mathematicians today than ever before; that 'nine times all previous' might well apply. The system of peer review is widespread and is ethically and mathematically reliable. Mathematics does not need the historic culture, but it firmly believes that the culture is an intrinsic part of the mathematics.

It isn't. The desire to make evaluations of worth within the discipline is unnecessary. It leads to restricted thinking and blindness to new direction. The rejection of all outside influence within the discipline reinforces its extreme insularity and its frequent detachment from even its own reality. That all of these cultural positions are held with a self-righteousness that is haughty and disdainful, carries the tenor of those who ordered society in earlier centuries.

It is not enough to have donned a T-shirt and jeans appearance if the iron hand can still crash on the desk with a resounding, *"Because I say so!"*

The analysis I have given here shows that Corpus Mathematica is behaving as would be expected. The corpus needs to understand that there are cultural imperatives operating from outside their own organisation. These are the same as for every other institution in society. If Corpus Mathematica continues with its exclusionist, ordering, and authoritarian culture, then at some time this will be seen for what it is. The corpus will then either die or undergo a forced and extremely painful reform. It is deluded in believing this inviolability will go on for ever. The cliff-edge may be very near.

We all need our history. We need to know it and to learn from it. What we cannot afford to do is hug history too close, with a heavy emotional investment. Far too many of the world's problems are based in that stultifying mind-set. My own belief is that many of those who formed our history are up there begging us to stop holding all that past as our only guide to the future.

The dangers of mathematics and its culture remaining tied together are enormous. Past mathematicians gave much of value, but they were human. They did not, could not, know everything. Putting all of them and their contributions together, it would be utter folly to say that this encapsulates all possible mathematical discovery. There have been some valuable new directions but consider how much still runs in the grooves of study gouged out centuries ago. Are there are so few other possibilities of direction, or if there are, that these could only be 'trivial'?

As a first step to self–reform (unlikely, but one should never give up hope) mathematicians need to stop talking to each other and listen to those outside. People have their own innate number sense and they have their own good common sense. When they were pupils, many could not wait to get out of the maths classroom. You, the mathematicians, glided over the inconsistencies and confusions in order to enter the discipline. For many others, these and an atmosphere that was authoritarian, was more than enough reason to walk away. You can go on telling each other that most walk away because it is a difficult subject. It is that and it always will be, but you are still kidding yourselves. A lot of people, when they were young, saw straight through you, and found themselves looking into a backward facing, rigid, and deeply authoritarian culture.

And so, in conclusion....

I can never believe that I am the first human being to have seen The Triples Generator. At the least one of the ancients must have seen it, even if his record is lost. Another good reason for using an 'on the tin what's in the tin' name.

It may be that The Triples Generator and the possibilities with the quartics and the quintics don't amount to a row of beans. But without finding these series and investigating them, one can never know. And infinite division matrices are not that thick on the ground.

What should not happen is to make an obsession out of triples, quadruples, or anything else for that matter. In view of the discovery, if it be that, coming from outside mathematics the chances of serious investigation are slim. Under the corpus's current cultural laws, even study, let alone obsession, will not be allowed to happen.

I make no apology for my presentation being unmathematical. Just so that nobody is in any doubt - no apology whatsoever.

I hope that my description of The Triples Generator is helpful to those who are interested but are themselves, like me, largely unmathematical. If it hasn't got through for you then I do apologise to you and to you alone. All I can say is, I've tried the best I can. Perhaps when you move away from the book, more will become apparent; as I said earlier, ideas are like that. I must thank you for your interest and for having stayed with me this far. At least you can take some comfort that, in all likelihood, you have been on exactly the same page as some of the most brilliant minds on the planet.

Authors are normally expected to put there acknowlegements on a separate page. For this book, they are just too integral for such a separation. Only one person can be named, my wife, and she is where she should be – at the front of this book. Without her love, and support it would never have happened. The usual 'responsibility' thing.

I wish to make it clear that no–one else will be named because the issues raised in this book are not about individuals. I cannot avoid my own individuality in having, apparently, discovered or re–discovered The Triples Generator, and chosen to make an analysis of culture in mathematics. But the question of individuals should stop there.

However, I must make an acknowledgement of special contributions. First, there were the attack dogs. But then it would be unfair to leave out the others: those who ignored my request for an interview, those who seemed to enjoy messing me around for an appointment, and those who rebuffed my request with rudeness. I should include those from earlier who ignored my request for guidance, sought because the mathematics books were so little help in what I was trying to understand. But then again, it's unfair to leave out those from much longer ago: the ones who taught me mathematics. For the most part they did it badly and with an all–pervading ill–temper.

I must thank you all. With your deeply–felt help I found The Triples Generator and my voice.

BONES 1: THE TRIPLES GENERATOR

Terms Used
The numerical values of the triple are:
- **Minor Value, Median Value, Major Value**.
- The Minor and Median Values can be generated in reverse order.

Therefore:
- **Addition Value 1 (AV1)**. This is the dividend value. The divisor will start to identify the triple by dividing on the square of this number. The square of this value is the one to be found in the line of differences.
- **Addition Value 2 (AV2)**. This is the number identified by processing AV1. The square of the AV2 is always in the line of squares.
- **Major Value**: its square is always in the line of squares.
- **Sin MiV**: the ratio tag for each triple. (If the triple were formed into a triangle this would be the sine of the minor angle.) Given by: Minor Value ÷ Major Value
- **Divisiors**: each of the numbers from 1 towards infinity. The divisors as a sequence form one axis of the infinite division matrix.
- **'Division by..'**: Refers to the series of divisors. Individual divisors are, for example, Division by 1 and are abbreviated to (÷1).
- **Dividend Values**. The number sequence from 3 onwards. These form the other axis of the infinite division matrix.
- **Division Interval**: The step on which each divisor picks up its dividend values.

Forming a triple
The following process includes a slight short cut. Maths people should easily recognise it, though it makes no difference to the outcome.

1	Select a divisor.	
2	Find a dividend value.	This is **Addition Value 1, (AV1)**.
3	Square the dividend value.	
4	Divide the square.	
5	Subtract the divisor.	
6	Divide by 2.	This is **Addition Value 2, (AV2)**.
7	Add the divisor.	This is the **Major Value**.

Parameters for forming a triple
- A triple cannot be formed by a divisor dividing on its own value.
- An even divisor must divide on an even dividend value.
- An odd divisor must divide on an odd dividend value.
- No triple can be formed on a dividend value < 3.

Division Interval

This is the critical parameter and must be dealt with separately. <u>First point</u>: I have not been able to identify a system, or an algorithm, or anything, really, that can be used to give the division interval in all cases.

Table 3.2 shows how the new ratio divisors are produced, in the early stages, in ($\div$1) and ($\div$2). I have shown the division interval each new ratio divisor will use and how that is a proportion of each divisor.

I can give the following:
- New ratio divisors. They are either an odd number square or 2×Square. The division interval is 2×The Root, in both cases.
- All other divisors multiply an earlier new ratio series. The multiplying factor must also multiply the division interval of the originating series.
- An even divisor multiplying up ($\div$2) will use its own value as its interval.
- An odd divisor multiplying up ($\div$1) will use 2× its own value as its interval.

I can go on shifting this around and get nowhere. I'm going to have to hand it over, that is, if there is anyone prepared to receive it.

What I believe to be the important characteristics of the generator have been covered in Chapters 1 to 4.

BISCUITS: A SUMMARY OF THE TABLES.

Table 1: The start of The Triples Generator.
    Dividend values 3 – 60. Divisors 1 – 18.

Tables at 2: The x^2 series 0 to 125. Triples placed in the series.
    Table 2.1: Division by 1.
    Table 2.2: Division by 2.
    Table 2.3: Division by 8.
    Table 2.4: Division by 9.
    Table 2.5: The x^2 series, range 304 to 425; the triple 297,304,425.
    (8th in Division by 121)

Tables at 3: The New Ratio Divisors. (First 25 triples)
    Table 3.1: Division by 1 and Division by 2.
    Table 3.2: The Maj/Min differences from (÷1) and (÷2).
    Table 3.3: Division by 8 and Division by 9.
    Table 3.4: Division by 18 and Division by 25.
    Table 3.5: Division by 32 and Division by 49.
    Table 3.6: Division by 50 and Division by 72.
    Table 3.7: Division by 81 and Division by 98.
    Table 3.8: Division by 1009 and Division by 1018081
        Note: (÷1009), prime, multiplying divisor.
        (÷1018081) is 1009^2, new ratio divisor.
    Table 3.9: Primes in the Major Values.

Tables 4.1 to 4.8: The Multiplying Divisors. (÷3) to (÷83).

Table 5: The Reverse Triples. Produced in normal form and found in reverse form in (÷1) to (÷83), and in (÷98).

Tables Summary: Some notes.
    The shade coding used in Tables 2.1 to 2.5 should make it easier to see and understand how a triple exists in the x^2 series. In Tables 3 and 4, and in Table 5, the shade coding shows both those triples which are actually shown in Tables 2.1 to 2.4 and those which can be found in Tables 2.1 to 2.4 (x range: 0 to 125) or in Table 2.5 (x range: 304 to 425).

It might be helpful if I indicated how to set about identifying some, one, of these for yourself. (Please don't say you were thinking of trying to do the lot!)

For those in the range of x: 0 to 125 it would be best to use Table 2.1 as this is the least cluttered with existing triples.
- Identify the triple you want to find and note the divisor.
- Find the second addition value (AV2) in the line of x. Immediately below this is the square of the value.
- Mark, or hold your finger on this point.
- Start adding up the differences from here on. The differences are the 3rd line of the set, marked $2x + 1$.
- When you've added on the last difference before the Major Value of your triple you've completed the summation part.
- For checking you might want to make a note of that summation total, or bung it into memory.
- Find the square root. That is the first addition value (AV1).
- You stopped with the last difference before the Major Value, so now you've identified the triple.

For checking, there are a couple of things you can do other than the obvious of: $\sqrt{((AV1)^2 + (AV2)^2)} = $ Major Value.

- Count the number of differences that you summed. This will be the same as the divisor. (Divisor = Major Value – AV2)
- Add the differences from $x = 0$, the first is 1, and keep adding until you reach the second addition value. That will be $(AV2)^2$. Add this to memory (Step 6 above), take the square root; you now have the Major Value.

If you have done all of this, finding the triple and then checking, it might have helped you to understand how the x^2 series works. It really is a summation series every bit as much as it is a multiplication series. Whether you've done it or just read what's here, that is more than enough. It really is time you went and met your friends. Have a nice evening.

THE TABLES

Table 1 (1 of 2)									The start of The
Dividend Values	Division by 1	Division by 2	Division by 3	Division by 4	Division by 5	Division by 6	Division by 7	Division by 8	Division by 9
3	3,4,5								
4		4,3,5							
5	5,12,13								
6		6,8,10							
7	7,24,25								
8		8,15,17		8,6,10					
9	9,40,41		9,12,15						
10		10,24,26							
11	11,60,61								
12		12,35,37		12,16,20		12,9,15		12,5,13	
13	13,84,85								
14		14,48,50							
15	15,112,113		15,36,39		15,20,25				15,8,17
16		16,63,65		16,30,34				16,12,20	
17	17,144,145								
18		18,80,82				18,24,30			
19	19,180,181								
20		20,99,101		20,48,52				20,21,29	
21	21,220,221		21,72,75				21,28,35		21,20,29
22		22,120,122							
23	23,264,265								
24		24,143,145		24,70,74		24,45,51		24,32,40	
25	25,312,313				25,60,65				
26		26,168,170							
27	27,364,365		27,120,123						27,36,45
28		28,195,197		28,96,100				28,45,53	
29	29,420,421								
30		30,224,226				30,72,78			
31	31,480,481								
32		32,255,257		32,126,130				32,60,68	
33	33,544,545		33,180,183						33,56,65
34		34,288,290							
35	35,612,613				35,120,125		35,84,91		
36		36,323,325		36,160,164		36,105,111		36,77,85	
37	37,684,685								
38		38,360,362							
39	39,760,761		39,252,255						39,80,89
40		40,399,401		40,198,202				40,96,104	
41	41,840,841								
42		42,440,442				42,144,150			
43	43,924,925								
44		44,483,485		44,240,244				44,117,125	
45	45;1012;1013		45,336,339		45,200,205				45,108,117
46		46,528,530							
47	47;1104;1105								
48		48,575,577		48,286,290		48,189,195		48,140,148	
49	49;1200;1201						49,168,175		
50		50,624,626							
51	51;1300;1301		51,432,435						51,140,149
52		52,675,677		52,336,340				52,165,173	
53	53;1404;1405								
54		54,728,730				54,240,246			
55	55;1512;1513				55,300,305				
56		56,783,785		56,390,394				56,192,200	
57	57;1624;1625		57,108,111						57,176,185
58		58,840,842							
59	59;1740;1741								
60		60,899,901		60,448,452		60,297,303		60,221,229	

Shown in Tables 2.1 to 2.4 Range x = 0 to 125
These can be found using the range x = 0 to 125, Tables 2.1 to 2.4
These can be found using the range x = 304 to 425, Table 2.5

Triples Generator   (2 of 2)   Table 1

Division by 10	Division by 11	Division by 12	Division by 13	Division by 14	Division by 15	Division by 16	Division by 17	Division by 18
20,15,25								
		24,18,30				24,10,26		24,7,25
				28,21,35				
30,40,50								30,16,34
						32,24,30		
	33,44,55							
		36,48,60						36,27,45
			39,52,65					
40,75,85						40,42,58		
				42,56,70				42,40,58
					45,60,75			
		48,90,102				48,64,80		48,55,73
50,120,130								
							51,68,85	
								54,72,90
	55,132,143							
				56,105,119		56,90,106		
60,175,185								60,91,109

Shown in Tables 2.1 to 2.4   Range x = 0 to 125
These can be found using the range x = 0 to 125, Tables 2.1 to 2.4
These can be found using the range x = 304 to 425, Table 2.5

Table 2.1 (1 of 2)   **Table to show triples arising from**

x	0		1		2		3		4		5
x²	0		1		4		9		16		25
2x+1		1		3		5		7		9	
√Σ(2x+1)										3	
(Line 1) →											→

x	11		12		13		14		15		16
x²	121		144		169		196		225		256
2x+1		23		25		27		29		31	
√Σ(2x+1)				5							
(Line 2) →											→

x	22		23		24		25		26		27
x²	484		529		576		625		676		729
2x+1		45		47		49		51		53	
√Σ(2x+1)						7					
(Line 3) →											→

x	33		34		35		36		37		38
x²	1089		1156		1225		1296		1369		1444
2x+1		67		69		71		73		75	
√Σ(2x+1)											
(Line 4) →											→

x	44		45		46		47		48		49
x²	1936		2025		2116		2209		2304		2401
2x+1		89		91		93		95		97	
√Σ(2x+1)											
(Line 5) →											→

x	55		56		57		58		59		60
x²	3025		3136		3249		3364		3481		3600
2x+1		111		113		115		117		119	
√Σ(2x+1)											
(Line 6) →											→

x	66		67		68		69		70		71
x²	4356		4489		4624		4761		4900		5041
2x+1		133		135		137		139		141	
√Σ(2x+1)											
(Line 7) →											→

x	77		78		79		80		81		82
x²	5929		6084		6241		6400		6561		6724
2x+1		155		157		159		161		163	
√Σ(2x+1)											
(Line 8) →											→

x	88		89		90		91		92		93
x²	7744		7921		8100		8281		8464		8649
2x+1		177		179		181		183		185	
√Σ(2x+1)											
(Line 9) →											→

x	99		100		101		102		103		104
x²	9801		10000		10201		10404		10609		10816
2x+1		199		201		203		205		207	
√(2x+1)											
(Line10) →											→

x	110		111		112		113		114		115
x²	12100		12321		12544		12769		12996		13225
2x+1		221		223		225		227		229	
√Σ(2x+1)						15					
(Line11) →											→

x	121		122		123		124		125	
x²	14641		14884		15129		15376		15625	
2x+1		243		245		247		249		251
√Σ(2x+1)										
(Line 12) →									→ ∞	

Diffs		AV1		AV2		Maj Val	

Division by 1. Range: x = 0 to x = 125 — (2 of 2) Table2.1

	6		7		8		9		10	
	36		49		64		81		100	
11		13		15		17		19		21
→										
	17		18		19		20		21	
	289		324		361		400		441	
33		35		37		39		41		43
→										
	28		29		30		31		32	
	784		841		900		961		1024	
55		57		59		61		63		65
→										
	39		**40**		**41**		42		43	
	1521		1600		1681		1764		1849	
77		79		**81**		83		85		87
				9						
→										
	50		51		52		53		54	
	2500		2601		2704		2809		2916	
99		101		103		105		107		109
→										
	61		62		63		64		65	
	3721		3844		3969		4096		4225	
121		123		125		127		129		131
11										
→										
	72		73		74		75		76	
	5184		5329		5476		5625		5776	
143		145		147		149		151		153
→										
	83		**84**		**85**		86		87	
	6889		7056		7225		7396		7569	
165		167		**169**		171		173		175
				13						
→										
	94		95		96		97		98	
	8836		9025		9216		9409		9604	
187		189		191		193		195		197
→										
	105		106		107		108		109	
	11025		11236		11449		11664		11881	
209		211		213		215		217		219
→										
	116		117		118		119		120	
	13456		13689		13924		14161		14400	
231		233		235		237		239		241
→										

Diffs	AV1	AV2	Maj Val

Table2.2 (1 of 2)   **Table to show triples arising from**

x	0		1		2		3		4		5
x²	0		1		4		9		16		25
2x+1		1		3		5		7		9	
√Σ(2x+1)								4			
(Line 1) →											→

x	11		12		13		14		15		16
x²	121		144		169		196		225		256
2x+1		23		25		27		29		31	
√Σ(2x+1)											8
(Line 2) →											→

x	22		23		24		25		26		27
x²	484		529		576		625		676		729
2x+1		45		47		49		51		53	
√Σ(2x+1)						10					
(Line 3) →											→

x	33		34		35		36		37		38
x²	1089		1156		1225		1296		1369		1444
2x+1		67		69		71		73		75	
√Σ(2x+1)						12					
(Line 4) →											→

x	44		45		46		47		48		49
x²	1936		2025		2116		2209		2304		2401
2x+1		89		91		93		95		97	
√Σ(2x+1)											14
(Line 5) →											→

x	55		56		57		58		59		60
x²	3025		3136		3249		3364		3481		3600
2x+1		111		113		115		117		119	
√Σ(2x+1)											
(Line 6) →											→

x	66		67		68		69		70		71
x²	4356		4489		4624		4761		4900		5041
2x+1		133		135		137		139		141	
√Σ(2x+1)											
(Line 7) →											→

x	77		78		79		80		81		82
x²	5929		6084		6241		6400		6561		6724
2x+1		155		157		159		161		163	
√Σ(2x+1)								18			
(Line 8) →											→

x	88		89		90		91		92		93
x²	7744		7921		8100		8281		8464		8649
2x+1		177		179		181		183		185	
√Σ(2x+1)											
(Line 9) →											→

x	99		100		101		102		103		104
x²	9801		10000		10201		10404		10609		10816
2x+1		199		201		203		205		207	
√(2x+1)			20								
(Line10) →											→

x	110		111		112		113		114		115
x²	12100		12321		12544		12769		12996		13225
2x+1		221		223		225		227		229	
√Σ(2x+1)											
(Line11) →											→

x	121		122		123		124		125		
x²	14641		14884		15129		15376		15625		
2x+1		243		245		247		249		251	
√Σ(2x+1)	22										
(Line 12) →										→ ∞	

Diffs	AV1	AV2	Maj Val

Division by 2. Range: x = 0 to x = 125

6	7	**8**	9	**10**
96	49	64	81	100
11	13	15	**17**	**19**	21
			6		
17	18	19	20	21
289	324	361	400	441
33	35	37	39	41	43
28	29	30	31	32
784	841	900	961	1024
55	57	59	61	63	65
39	40	41	42	43
1521	1600	1681	1764	1849
77	79	81	83	85	87
50	51	52	53	54
2500	2601	2704	2809	2916
99	101	103	105	107	109
61	62	**63**	64	**65**
3721	3844	3969	4096	4225
121	123	125	**127**	**129**	131
			16		
72	73	74	75	76
5184	5329	5476	5625	5776
143	145	147	149	151	153
83	84	85	86	87
6889	7056	7225	7396	7569
165	167	169	171	173	175
94	95	96	97	98
8836	9025	9216	9409	9604
187	189	191	193	195	197
105	106	107	108	109
11025	11236	11449	11664	11881
209	211	213	215	217	219
116	117	118	119	**120**
13456	13689	13924	14161	14400
231	233	235	237	239	**241**

Diffs	AV1	AV2	Maj Val

Table 2.3 (1 of 2)   **Table to show triples arising from**

Line 1

x	0		1		2		3		4	5
x²	0		1		4		9		16	25
2x+1		1		3		5		7		9
√Σ(2x+1)										
(Line 1)	→									→

Line 2

x	11		12		13		14		15	16
x²	121		144		169		196		225	256
2x+1		23		25		27		29		31
√Σ(2x+1)										16
(Line 2)	→									→

Line 3

x	22		23		24		25		26	27
x²	484		529		576		625		676	729
2x+1		45		47		49		51		53
√Σ(2x+1)							20			
(Line 3)	→									→

Line 4

x	33		34		35		36		37	38
x²	1089		1156		1225		1296		1369	1444
2x+1		67		69		71		73		75
√Σ(2x+1)							24			
(Line 4)	→									→

Line 5

x	44		45		46		47		48	49
x²	1936		2025		2116		2209		2304	2401
2x+1		89		91		93		95		97
√Σ(2x+1)										28
(Line 5)	→									→

Line 6

x	55		56		57		58		59	60
x²	3025		3136		3249		3364		3481	3600
2x+1		111		113		115		117		119
√Σ(2x+1)										
(Line 6)	→									→

Line 7

x	66		67		68		69		70	71
x²	4356		4489		4624		4761		4900	5041
2x+1		133		135		137		139		141
√Σ(2x+1)										
(Line 7)	→									→

Line 8

x	77		78		79		80		81	82
x²	5929		6084		6241		6400		6561	6724
2x+1		155		157		159		161		163
√Σ(2x+1)									36	
(Line 8)	→									→

Line 9

x	88		89		90		91		92	93
x²	7744		7921		8100		8281		8464	8649
2x+1		177		179		181		183		185
√Σ(2x+1)										
(Line 9)	→									→

Line 10

x	99		100		101		102		103	104
x²	9801		10000		10201		10404		10609	10816
2x+1		199		201		203		205		207
√(2x+1)		40								
(Line10)	→									→

Line 11

x	110		111		112		113		114	115
x²	12100		12321		12544		12769		12996	13225
2x+1		221		223		225		227		229
√Σ(2x+1)										
(Line11)	→									→

Line 12

x	121		122		123		124		125	
x²	14641		14884		15129		15376		15625	
2x+1		243		245		247		249		251
√Σ(2x+1)	44									
(Line 12)	→								→ ∞	

Diffs	AV1	AV2	Maj Val

Note: The first two triples overlap. Both are reverse form.

Division by 8. Range: x = 0 to x = 125 (2 of 2) **Table 2.3**

	6	7	8	9	10		
	36	49	64	81	100		
11	**13**	**15**	**17**	**19**		**21**	
				12			
	17	18	19	**20**	**21**		
	289	324	361	400	441		
33	**35**	**37**	**39**	41		**43**	
	28	**29**	30	31	**32**		
	784	841	900	961	1024		
55	**57**	59	61	63		**65**	
	39	**40**	41	42	43		
	1521	1600	1681	1764	1849		
77	**79**	81	83	85		87	
	50	51	52	**53**	54		
	2500	2601	2704	2809	2916		
99	**101**	**103**	**105**	107		109	
	61	62	63	64	65		
	3721	3844	3969	4096	4225		
121	**123**	**125**	**127**	**129**		**131**	
				32			
	72	73	74	75	76		
	5184	5329	5476	5625	5776		
143	145	147	149	151		153	
	83	84	**85**	86	87		
	6889	7056	7225	7396	7569		
165	**167**	**169**	171	173		175	
	94	95	**96**	97	98		
	8836	9025	9216	9409	9604		
187	189	191	**193**	**195**		**197**	
	105	106	107	108	109		
	11025	11236	11449	11664	11881		
209	211	213	215	217		219	
	116	**117**	118	119	120		
	13456	13689	13924	14161	14400		
231	233	**235**	**237**	**239**		**241**	

Diffs	AV1	AV2	Maj Val

Table 2.4 (1 of 2)   **Table to show triples arising from**

x	0	1	2	3	4	5
x²	0	1	4	9	16	25
2x+1		1	3	5	7	9
√Σ(2x+1)						
(Line 1)	→					→

x	11	12	13	14	15	16
x²	121	144	169	196	225	256
2x+1	**23**		**25**	**27**	**29**	**31**
√Σ(2x+1)			**15**			
(Line 2)	→					→

x	22	23	24	25	26	27
x²	484	529	576	625	676	729
2x+1	**45**		**47**	**49**	**51**	**53**
√Σ(2x+1)				**21**		
(Line 3)	→					→

x	33	34	35	**36**	37	38
x²	1089	1156	1225	1296	1369	1444
2x+1	67	69	71		**73**	**75**
√Σ(2x+1)						
(Line 4)	→					→

x	44	**45**	46	47	48	49
x²	1936	2025	2116	2209	2304	2401
2x+1	**89**		91	93	95	97
√Σ(2x+1)						
(Line 5)	→					→

x	55	**56**	57	58	59	60
x²	3025	3136	3249	3364	3481	3600
2x+1	111		**113**	**115**	**117**	**119**
√Σ(2x+1)						
(Line 6)	→					→

x	66	67	68	69	70	71
x²	4356	4489	4624	4761	4900	5041
2x+1	133	135	137	139	141	
√Σ(2x+1)						
(Line 7)	→					→

x	77	78	79	**80**	81	82
x²	5929	6084	6241	6400	6561	6724
2x+1	155	157	159		**161**	**163**
√Σ(2x+1)						
(Line 8)	→					→

x	88	**89**	90	91	92	93
x²	7744	7921	8100	8281	8464	8649
2x+1	**177**		179	181	183	185
√Σ(2x+1)						
(Line 9)	→					→

x	99	100	101	102	103	104
x²	9801	10000	10201	10404	10609	10816
2x+1	199	201	203	205	207	
√(2x+1)						
(Line10)	→					→

x	110	111	112	113	114	115
x²	12100	12321	12544	12769	12996	13225
2x+1	**221**		**223**	**225**	**227**	**229**
√Σ(2x+1)				**45**		
(Line11)	→					→

x	121	122	123	124	125	
x²	14641	14884	15129	15376	15625	
2x+1	243	245	247	249	251	
√Σ(2x+1)						
(Line 12)	→				→ ∞	

Diffs	AV1	AV2	Maj Val

Division by 9. Range: x = 0 to x = 125   (2 of 2) Table 2.4

Col1	Col2	Col3	Col4	Col5	Col6
	6	7	**8**	9	10
	36	49	64	81	100
11	13	15	**17**	**19**	**21**
→					
	17	18	19	**20**	21
	289	324	361	400	441
33	35	37	39	**41**	**43**
→					
	28	**29**	30	31	32
	784	841	900	961	1024
55	**57**	59	61	63	65
→					
	39	40	41	42	43
	1521	1600	1681	1764	1849
77	**79**	**81**	83	85	87
		27			
→					
	50	51	52	53	54
	2500	2601	2704	2809	2916
99	101	103	105	107	109
→					
	61	62	63	64	**65**
	3721	3844	3969	4096	4225
121	**123**	**125**	**127**	**129**	131
→					
	72	73	74	75	76
	5184	5329	5476	5625	5776
143	145	147	149	151	153
→					
	83	84	85	86	87
	6889	7056	7225	7396	7569
165	**167**	**169**	**171**	**173**	**175**
		39			
→					
	94	95	96	97	98
	8836	9025	9216	9409	9604
187	189	191	193	195	197
→					
	105	106	107	**108**	109
	11025	11236	11449	11664	11881
209	211	213	215	**217**	**219**
→					
	116	**117**	118	119	120
	13456	13689	13924	14161	14400
231	**233**	235	237	239	241
→					

Diffs	AV1	AV2	Maj Val

Table 2.5 (1 of 2)   **The triple 297,304,425 at its position in the x² series. Generated**

x	From zero →	**304**	305	306	307	308
x²		92416	93025	93636	94249	94864
2x+1		609	611	613	615	
√∑(2x+1)						
(Line 1) →						→
x	314	315	316	317	318	319
x²	98596	99225	99856	100489	101124	101761
2x+1	629	631	633	635	637	
√∑(2x+1)						
(Line 2) →						→
x	325	326	327	328	329	330
x²	105625	106276	106929	107584	108241	108900
2x+1	651	653	655	657	659	
√∑(2x+1)						
(Line 3) →						→
x	336	337	338	339	340	341
x²	112896	113569	114224	114921	115600	116281
2x+1	673	675	677	679	681	
√∑(2x+1)						
(Line 4) →						→
x	347	348	349	350	351	352
x²	120409	121104	121801	122500	123201	123904
2x+1	695	697	699	701	703	
√∑(2x+1)						
(Line 5) →						→
x	358	359	360	361	362	363
x²	128164	128881	129600	130321	131044	131769
2x+1	717	719	721	723	725	
√∑(2x+1)						
(Line 6) →						→
x	369	370	371	372	373	374
x²	136161	136900	137641	138384	139129	139876
2x+1	739	741	743	745	747	
√∑(2x+1)						
(Line 7) →						→
x	380	381	382	383	384	385
x²	144400	145161	145924	146689	147456	148225
2x+1	761	763	765	767	769	
√∑(2x+1)						
(Line 8) →						→
x	391	392	393	394	395	396
x²	152881	153664	154449	155236	156025	156816
2x+1	783	785	787	789	791	
√∑(2x+1)						
(Line 9) →						→
x	402	403	404	405	406	407
x²	161604	162409	163216	164025	164836	165649
2x+1	805	807	809	811	813	
√(2x+1)						
(Line10) →						→
x	413	414	415	416	417	418
x²	170569	171396	17225	173056	173889	174724
2x+1	827	829	831	833	835	
√∑(2x+1)						
(Line11) →						→
x	424	**425**				
x²	179776	180625	→ ∞			
2x+1	849					
√∑(2x+1)						
(Line12) →						
	Diffs		**AV1**		**AV2**	**Maj Val**

in the Division by 121 series. Range: x = 304 to x = 425   (2 of 2) Table 2.5

	309	310	311	312	313	
	95481	96100	96721	97344	97969	
617	619	621	623	625		627
→						
	320	321	322	323	324	
	102400	103041	103684	104329	104976	
639	641	643	645	647		649
→						
	331	332	333	334	335	
	109561	110224	110889	111556	112225	
661	663	665	667	669		671
→						
	342	343	344	345	346	
	116964	117649	118336	119025	119716	
683	685	687	689	691		693
→						
	353	354	355	356	357	
	124609	125316	126025	126736	127449	
705	707	709	711	713		715
→						
	364	365	366	367	368	
	132496	133225	133956	134689	135424	
727	729	731	733	735		737
	297					
→						
	375	376	377	378	379	
	140625	141376	142129	142884	143641	
749	751	753	755	757		759
→						
	386	387	388	389	390	
	148996	149769	150544	151321	152100	
771	773	775	777	779		781
→						
	397	398	399	400	401	
	157609	158404	159201	160000	160801	
793	795	797	799	801		803
→						
	408	409	410	411	412	
	166464	167281	168100	168921	169744	
815	817	819	821	823		825
→						
	419	420	421	422	423	
	175561	176400	177241	178084	178929	
837	839	841	843	845		847
→						
	Diffs		AV1		AV2	Maj Val

Table 3.1 — **New Ratio Divisors**

Division by 1			Division interval:2		Division by 2			Division interval: 2	
AV 1	AV 2	Maj. Val.	Sin MiV	(code)	AV1	AV2	Maj. Val.	Sin MiV	(code)
3	4	5	0.6	x p	4	3	5	0.6	1.p
5	12	13	0.3846	x p	6	8	10	0.6	1
7	24	25	0.28	x	8	15	17	0.4706	x p
9	40	41	0.2195	x p	10	24	26	0.3846	1
11	60	61	0.1803	x p	12	35	37	0.3243	x p
13	84	85	0.1529	x	14	48	50	0.28	1
15	112	113	0.1327	x p	16	63	65	0.2462	x
17	144	145	0.1172	x	18	80	82	0.2195	1
19	180	181	0.105	x p	20	99	101	0.198	x p
21	220	221	0.095	x	22	120	122	0.1803	1
23	264	265	0.0868	x	24	143	145	0.1655	x
25	312	313	0.0799	x p	26	168	170	0.1529	1
27	364	365	0.074	x	28	195	197	0.1421	x p
29	420	421	0.0689	x p	30	224	226	0.1327	1
31	480	481	0.0644	x	32	255	257	0.1245	x p
33	544	545	0.0606	x	34	288	290	0.1172	1
35	612	613	0.0571	x p	36	323	325	0.1108	x
37	684	685	0.054	x	38	360	362	0.105	1
39	760	761	0.0512	x p	40	399	401	0.0998	x p
41	840	841	0.0488	x	42	440	442	0.095	1
43	924	925	0.0465	x	44	483	485	0.0907	x
45	1012	1013	0.0444	x p	46	528	530	0.0868	1
47	1104	1105	0.0425	x	48	575	577	0.0832	x p
49	1200	1201	0.0408	x p	50	624	626	0.0799	1
51	1300	1301	0.0392	x p	52	675	677	0.0768	x p

Code:

x – A new ratio of triple

p – Prime number in the major value

Numbers: Above the dotted line: The number is the series from which the reverse triple was drawn.

Below the dotted line: The number is the series where that ratio first arose.

Shown in Tables 2.1 and 2.2

Can be found using Table 2.5

Table 3.2 — **New Ratio Divisors**

Difference between minor and major values in Division by 1 and Division by 2 series.

From Division by 1 series				From Division by 2 series				
(A) Minor Value	(B) Major Value	New ratio divisors (B) − (A)	Div Int., & its prop'n of divisor	(A) Minor Value	(B) Major Value	Divisor given by: (B) − (A)	New ratios/ Multipliers	Div Int., & its prop'n of Divisor
3	5	2	2 (1/1)	Reverse triple: 4,3,5; diff. is not applicable.				n/a
5	13	8	4 (1/2)	6	10	4	→ 2×(÷2)	4 (1/1)
7	25	18	6 (1/3) x	8	17	9	New ratios	6 (2/3)
9	41	32	8 (1/4)	10	26	16	→ 2×(÷8)	8 (1/2)
11	61	50	10 (1/5) x	12	37	25	New ratios	10 (2/5)
13	85	72	12 (1/6)	14	50	36	→ 2×(÷18)	12 (1/3)
15	113	98	14 (1/7) x	16	65	49	New ratios	14 (2/7)
17	145	128	16 (1/8)	18	82	64	→ 2×(÷32)	16 (1/4)
19	181	162	18 (1/9) x	20	101	81	New ratios	18 (2/9)
21	221	200	20 (1/10)	22	122	100	→ 2×(÷50)	20 (1/5)
23	265	242	22 (1/11) x	24	145	121	New ratios	22 (2/11)
25	313	288	24 (1/12)	26	170	144	→ 2×(÷72)	24 (1/6)
27	365	338	26 (1/13) x	28	197	169	New ratios	26 (2/13)
29	421	392	28 (1/14)	30	226	196	→2×(÷98)	28 (1/7)
31	481	450	30 (1/15) x	32	257	225	New ratios	30 (2/15)
33	545	512	32 (1/16)	34	290	256	→2×(÷128)	32 (1/8)
35	613	578	34 (1/17) x	36	325	289	New ratios	34 (2/17)
37	685	648	36 (1/18)	38	362	324	→2×(÷162)	36 (1/9)
39	761	722	38 (1/19) x	40	401	361	New ratios	38 (2/19)
41	841	800	40 (1/20)	42	442	400	→2×(÷200)	40 (1/10)
43	925	882	42 (1/21) x	44	485	441	New ratios	42 (2/21)
45	1013	968	44 (1/22)	46	530	484	→2×(÷242)	44 (1/11)
47	1105	1058	46 (1/23) x	48	577	529	New ratios	46 (2/23)
49	1201	1152	48 (1/24)	50	626	576	→2×(÷288)	48 (1/12)
51	1301	1250	50 (1/25) x	52	677	625	New ratios	50 (2/25)

Notes

Division by 1: Major/Minor difference is the sequence of (2 × series of squares);

Division by 2 produces the sequence of squares.

All Division by 1 values become series which contain new ratios. From Division by 2, only the odd number squares become divisors whose series contain new ratios.

All new ratio divisors bring forward their own triple from either (÷1) or (÷2) as their first and reverse, triple. Subsequent reverse triples come from series earlier than the divisor in question. Divisor = Maj/Min difference. See Chapter 4.

The maj/min difference continues across the new ratio series, at 2 × square in the odd new ratio divisors, and the square in the even new ratio divisors. It seems very likely that only (÷1) and (÷2) series will produce new ratio divisors.

Table 3.3 — **New Ratio Divisors**

Division by 8			Division interval: 4		Division by 9			Division interval: 6	
AV1	AV2	Maj. Val.	Sin MiV	(code)	AV1	AV2	Maj. Val.	Sin MiV	(code)
12	5	13	0.3846	1.p	15	8	17	0.4706	2
16	12	20	0.6	4	21	20	29	0.6897	8.p
20	21	29	0.6897	x p	27	36	45	0.6	1
24	32	40	0.6	1	33	56	65	0.5077	x
28	45	53	0.5283	x p	39	80	89	0.4382	x p
32	60	68	0.4706	2	45	108	117	0.3846	1
36	77	85	0.4235	x	51	140	149	0.3423	x p
40	96	104	0.3846	1	57	176	185	0.3081	x
44	117	125	0.352	x	63	216	225	0.28	1
48	140	148	0.3243	2	69	260	269	0.2565	x p
52	165	173	0.3006	x p	75	308	317	0.2366	x p
56	192	200	0.28	1	81	360	369	0.2195	1
60	221	229	0.262	x p	87	416	425	0.2047	x
64	252	260	0.2462	2	93	476	485	0.1917	x
68	285	293	0.2321	x p	99	540	549	0.1803	1
72	320	328	0.2195	1	105	608	617	0.1702	x p
76	357	365	0.2082	x	111	680	689	0.1611	x
80	396	404	0.198	2	117	756	765	0.1529	1
84	437	445	0.1888	x	123	836	845	0.1456	x
88	480	488	0.1803	1	129	920	929	0.1389	x p
92	525	533	0.1726	x	135	1008	1017	0.1327	1
96	572	580	0.1655	2	141	1100	1109	0.1271	x p
100	621	629	0.159	x	147	1196	1205	0.122	x
104	672	680	0.1529	1	153	1296	1305	0.1172	1
108	725	733	0.1473	x p	159	1400	1409	0.1128	x p

Code:

x – A new ratio of triple

p – Prime number in the major value

Numbers: Above the dotted line: The number is the series from which the reverse triple was drawn.

Below the dotted line: The number is the series where that ratio first arose.

Shown in Tables 2.3 and 2.4

Can be found using Table 2.5

Table 3.4				**New Ratio Divisors**					
Division by 18			Division interval: 6		**Division by 25**			Division interval: 10	
AV1	AV2	Maj. Val.	Sin MiV	(code)	AV1	AV2	Maj. Val	Sin MiV	(code)
24	7	25	0.28	1	35	12	37	0.3243	2.p
30	16	34	0.4706	4	45	28	53	0.5283	8.p
36	27	45	0.6	9	55	48	73	0.6575	18.p
42	40	58	0.6897	16	65	72	97	0.6701	x p
48	55	73	0.6575	x p	75	100	125	0.6	1
54	72	90	0.6	1	85	132	157	0.5414	x p
60	91	109	0.5505	x p	95	168	193	0.4922	x p
66	112	130	0.5077	9	105	208	233	0.4506	x p
72	135	153	0.4706	2	115	252	277	0.4152	x p
78	160	178	0.4382	9	125	300	325	0.3846	1
84	187	205	0.4098	x	135	352	377	0.3581	x
90	216	234	0.3846	1	145	408	433	0.3349	x p
96	247	265	0.3623	x	155	468	493	0.3144	x
102	280	298	0.3423	9	165	532	557	0.2962	x p
108	315	333	0.3243	2	175	600	625	0.28	1
114	352	370	0.3081	9	185	672	697	0.2654	x
120	391	409	0.2934	x p	195	748	773	0.2523	x p
126	432	450	0.28	1	205	828	853	0.2403	x p
132	475	493	0.2677	x	215	912	937	0.2295	x p
138	520	538	0.2565	9	225	1000	1025	0.2195	1
144	567	585	0.2462	2	235	1092	1117	0.2104	x p
150	616	634	0.2366	9	245	1188	1213	0.202	x p
156	667	685	0.2277	x	255	1288	1313	0.1942	x
162	720	738	0.2195	1	265	1392	1417	0.187	x
168	775	793	0.2118	x	275	1500	1525	0.1803	1

Code:

x – A new ratio of triple

p – Prime number in the major value

Numbers: Above the dotted line: The number is the series from which the reverse triple was drawn.

Below the dotted line: The number is the series where that ratio first arose.

Can be found using Tables 2.1 to 2.4

Can be found using Table 2.5

Table 3.5							New Ratio Divisors			
Division by 32			Division interval: 8		**Division by 49**			Division interval: 14		
AV1	AV2	Maj. Val.	Sin MiV	(code)	AV1	AV2	Maj. Val.	Sin MiV	(code)
40	9	41	0.2195	1.p	63	16	65	0.2462	2
48	20	52	0.3846	4	77	36	85	0.4235	8
56	33	65	0.5077	9	91	60	109	0.5505	18.p
64	48	80	0.6	16	105	88	137	0.6423	32.p
72	65	97	0.6701	25.p	119	120	169	0.7041	x
80	84	116	0.6897	8	133	156	205	0.6488	x
88	105	137	0.6423	x p	147	196	245	0.6	1
96	128	160	0.6	1	161	240	289	0.5571	x
104	153	185	0.5622	x	175	288	337	0.5193	x p
112	180	212	0.5283	8	189	340	389	0.4859	x p
120	209	241	0.4979	x p	203	396	445	0.4562	x
128	240	272	0.4706	2	217	456	505	0.4297	x
136	273	305	0.4459	x	231	520	569	0.406	x p
144	308	340	0.4235	8	245	588	637	0.3846	1
152	345	377	0.4032	x	259	660	709	0.3653	x p
160	384	416	0.3846	1	273	736	785	0.3477	x
168	425	457	0.3676	x p	287	816	865	0.3318	x
176	468	500	0.352	8	301	900	949	0.3172	x
184	513	545	0.3376	x	315	988	1037	0.3038	x
192	560	592	0.3243	2	329	1080	1129	0.2914	x p
200	609	641	0.312	x p	343	1176	1225	0.28	1
208	660	692	0.3006	8	357	1276	1325	0.2694	x
216	713	745	0.2899	x	371	1380	1429	0.2596	x p
224	768	800	0.28	1	385	1488	1537	0.2605	x
232	825	857	0.2707	x p	399	1600	1649	0.242	x

Code:

x – A new ratio of triple

p – Prime number in the major value

Numbers: Above the dotted line: The number is the series from which the reverse triple was drawn.

Below the dotted line: The number is the series where that ratio first arose.

Can be found using Tables 2.1 to 2.4

Can be found using Table 2.5

Table 3.6					New Ratio Divisors				
Division by 50			Division interval: 10		Division by 72			Division interval: 12	
AV1	AV2	Maj. Val	Sin MiV	(code)	AV1	AV2	Maj. Val.	Sin MiV	(code)
60	11	61	0.1803	1.p	84	13	85	0.1529	1
70	24	74	0.3243	4	96	28	100	0.28	4
80	39	89	0.4382	9.p	108	45	117	0.3846	9
90	56	106	0.5283	16	120	64	136	0.4706	16
100	75	125	0.6	25	132	85	157	0.5414	25.p
110	96	146	0.6575	36	144	108	180	0.6	36
120	119	169	0.7041	49	156	133	205	0.6488	49
130	144	194	0.6701	25	168	160	232	0.6897	64
140	171	221	0.6335	x	180	189	261	0.6897	8
150	200	250	0.6	1	192	220	292	0.6575	18
160	231	281	0.5694	x p	204	253	325	0.6277	x
170	264	314	0.5414	25	216	288	360	0.6	1
180	299	349	0.5158	x p	228	325	397	0.5743	x p
190	336	386	0.4922	25	240	364	436	0.5505	18
200	375	425	0.4706	2	252	405	477	0.5283	8
210	416	466	0.4506	25	264	448	520	0.5077	9
220	459	509	0.4322	x p	276	493	565	0.4885	x
230	504	554	0.4152	25	288	540	612	0.4706	2
240	551	601	0.3993	x p	300	589	661	0.4539	x p
250	600	650	0.3846	1	312	640	712	0.4382	9
260	651	701	0.3709	x p	324	693	765	0.4235	8
270	704	754	0.3581	25	336	748	820	0.4098	18
280	759	809	0.3461	x p	348	805	877	0.3968	x p
290	816	866	0.3349	25	360	864	936	0.3846	1
300	875	925	0.3243	2	372	925	997	0.3731	x p

Code:

x – A new ratio of triple

p – Prime number in the major value

Numbers: Above the dotted line: The number is the series from which the reverse triple was drawn.

Below the dotted line: The number is the series where that ratio first arose.

	Can be found using Tables 2.1 to 2.4
	Can be found using Table 2.5

Table 3.7				New Ratio Divisors					
Division by 81			Division interval: 18		**Division by 98**			Division interval: 14	
AV1	AV2	Maj. Val.	Sin MiV	(code)	AV1	AV2	Maj. Val.	Sin MiV	(code)
99	20	101	0.198	2.p	112	15	113	0.1327	1.p
117	44	125	0.352	8	126	32	130	0.2462	4
135	72	153	0.4706	18	140	51	149	0.3423	9.p
153	104	185	0.5622	32	154	72	170	0.4235	16
171	140	221	0.6335	50	168	95	193	0.4922	25.p
189	180	261	0.6897	72	182	120	218	0.5505	36
207	224	305	0.6787	x	196	147	245	0.6	49
225	272	353	0.6374	x p	210	176	274	0.6423	64
243	324	405	0.6	1	224	207	305	0.6787	81
261	380	461	0.5662	x p	238	240	338	0.7041	49
279	440	521	0.5355	x p	252	275	373	0.6756	x p
297	504	585	0.5077	9	266	312	410	0.6488	49
315	572	653	0.4824	x p	280	351	449	0.6236	x p
333	644	725	0.4593	x	294	392	490	0.6	1
351	720	801	0.4382	9	308	435	533	0.5779	x
369	800	881	0.4188	x p	322	480	578	0.5571	49
387	884	965	0.401	x	336	527	625	0.5376	x
405	972	1053	0.3846	1	350	576	674	0.5193	49
423	1064	1145	0.3694	x	364	627	725	0.5021	x
441	1160	1241	0.3553	x	378	680	778	0.4859	49
459	1260	1341	0.3423	2	392	735	833	0.4706	2
477	1364	1445	0.3301	x	406	792	890	0.4562	49
495	1472	1553	0.3187	x p	420	851	949	0.4426	x
513	1584	1665	0.3081	9	434	912	1010	0.4297	49
531	1700	1781	0.2981	x	448	975	1073	0.4175	x

Code:

x – A new ratio of triple

p – Prime number in the major value

Numbers: Above the dotted line: The number is the series from which the reverse triple was drawn.

Below the dotted line: The number is the series where that ratio first arose.

	Can be found using Tables 2.1 to 2.4
	Can be found using Table 2.5

Table 3.8					Prime Divisor 1009 and its Square 1018081				
Division by 1009			Division interval: 2018		**Division by 1018081**			Division interval: 2018	
AV1	AV2	Maj. Val.	Sin MiV	(code)*	AV1	AV2	Maj. Val.	Sin MiV	(code)
3027	4036	5045	0.6	2018	1020099	2020	1020101	0.00198	2
5045	12108	13117	0.3846	8072	1022117	4044	1022125	0.00396	8
7063	24216	25225	0.28	18162	1024135	6072	1024153	0.00593	18
9081	40360	41369	0.2195	32288	1026153	8104	1026185	0.0079	32
11099	60540	61549	0.1803	50450	1028171	10140	1028221	0.00986	50
13117	84756	85765	0.1529	72648	1030189	12180	1030261	0.01182	72
15135	113008	114017	0.1327	98882	1032207	14224	1032305	0.01378	98
17153	145296	146305	0.1172	129152	1034225	16272	1034353	0.01573	128
19171	181620	182629	0.105	163458	1036243	18324	1036405	0.01768	162
21189	221980	222989	0.095	201800	1038261	20380	1038461	0.01963	200
23207	266376	267385	0.0868	244178	1040279	22440	1040521	0.02157	242
25225	314808	315817	0.0799	290592	1042297	24504	1042585	0.0235	288
27243	367276	368285	0.074	341042	1044315	26572	1044653	0.02544	338
29261	423780	424789	0.0689	395528	1046333	28644	1046725	0.02737	392
31279	484320	485329	0.0644	454050	1048351	30720	1048801	0.02929	450
33297	548896	549905	0.0606	516608	1050369	32800	1050881	0.03121	512
35315	617508	618517	0.0571	583202	1052387	34884	1052965	0.03313	578
37333	690156	691165	0.054	653832	1054405	36972	1055053	0.03504	648
39351	766840	767849	0.0512	728498	1056423	39064	1057145	0.03695	722
41369	847560	848569	0.0488	807200	1058441	41160	1059241	0.03886	800
43387	932316	933325	0.0465	889938	1060459	43260	1061341	0.04076	882
45405	1021108	1022117	0.0444	976712	1062477	45364	1063445	0.04266	968
47423	1113936	1114945	0.0425	1067522	1064495	47472	1065553	0.04455	1058
49441	1210800	1211809	0.0408	1162368	1066513	49584	1067665	0.04644	1152
51459	1311700	1312709	0.0392	1261250	1068531	51700	1069781	0.04833	1250

Division by 1009: (÷1) series × 1009. (code)* shows the divisors which will reverse the triple. These divisors are 1009 × sequence shown in (code) for divisor 1018081.

Division by 1018081: First 25 triples shown. All are reverse triples. Another 688 to complete reverse sequence. Last of these will be drawn from Division by 1016738 (2 × 713²).

(code): identical to third column of Table 3.2, new ratio divisors generated in Division by 1.

Table 3.9	**New Ratio Divisors**					
Incidence of primes in the major values						
1	2	3	4	5	6	7
Division by... series	Number new ratio triples	No. primes Maj.Vals. in 2	3 as % of 2	Range in which 2 arise	Number primes in 5	6 as % of odds in 5
1	25	13	52%	5-1301	212	32.70%
2	12	8	66.66%	17-677	117	35.50%
8	12	6	50.00%	29-733	121	34.40%
9	15	8	53.30%	45-1409	209	30.60%
18	8	3	37.50%	73-793	118	32.80%
25	17	12	70.60%	97-1525	216	30.30%
32	10	5	50.00%	116-857	118	31.80%
49	18	6	33.33%	169-1649	220	29.70%
50	7	6	85.70%	194-925	114	31.20%
72	6	4	66.66%	261-997	113	30.70%
81	13	6	46.10%	305-1781	213	28.90%
98	7	2	28.60%	338-1073	112	30.50%
Totals	150	79	52.67%	5 - 1649	257	31.26%

Notes

Col 2: Does not include primes in the reverse triples. (They would be counted twice)

Col 7: Percentage calculated against odd numbers only to give a like comparison to Col 4.

Table 4.1											**The Multiplying Divisors**

Division by 3		DI:6	3 × (÷1)	**Division by 4**		DI:4	2 × (÷2)	**Division by 5**		DI:10	5 × (÷1)
AV1	AV2	Maj. Val.	Sin MiV	AV1	AV2	Maj. Val.	Sin MiV	AV1	AV2	Maj. Val.	Sin MiV
9	12	15	0.6	8	6	10	0.6	15	20	25	0.6
15	36	39	0.3846	12	16	20	0.6	25	60	65	0.3846
21	72	75	0.28	16	30	34	0.4706	35	120	125	0.28
27	120	123	0.2195	20	48	52	0.3846	45	200	205	0.2195
33	180	183	0.1803	24	70	74	0.3243	55	300	305	0.1803
39	252	255	0.1529	28	96	100	0.28	65	420	425	0.1529
45	336	339	0.1327	32	126	130	0.2462	75	560	565	0.1327
51	432	435	0.1172	36	160	164	0.2195	85	720	725	0.1172
57	540	543	0.105	40	198	202	0.198	95	900	905	0.105
63	660	663	0.095	44	240	244	0.1803	105	1100	1105	0.095

Division by 6		DI:6	3 × (÷2)	**Division by 7**		DI:14	7 × (÷1)	**Division by 10**		DI:10	5 × (÷2)
AV1	AV2	Maj. Val.	Sin MiV	AV1	AV2	Maj. Val.	Sin MiV	AV1	AV2	Maj. Val.	Sin MiV
12	9	15	0.6	21	28	35	0.6	20	15	25	0.6
18	24	30	0.6	35	84	91	0.3846	30	40	50	0.6
24	45	51	0.4706	49	168	175	0.28	40	75	85	0.4706
30	72	78	0.3846	63	280	287	0.2195	50	120	130	0.3846
36	105	111	0.3243	77	420	427	0.1803	60	175	185	0.3243
42	144	150	0.28	91	588	595	0.1529	70	240	250	0.28
48	189	195	0.2462	105	784	791	0.1327	80	315	325	0.2462
54	240	246	0.2195	119	1008	1015	0.1172	90	400	410	0.2195
60	297	303	0.198	133	1260	1267	0.105	100	495	505	0.198
66	360	366	0.1803	147	1540	1547	0.095	110	600	610	0.1803

Division by 11		DI:22	11 × (÷1)	**Division by 12**		DI:12	6 × (÷2)	**Division by 13**		DI:26	13 × (÷1)
AV1	AV2	Maj. Val.	Sin MiV	AV1	AV2	Maj. Val.	Sin MiV	AV1	AV2	Maj. Val.	Sin MiV
33	44	55	0.6	24	18	30	0.6	39	52	65	0.6
55	132	143	0.3846	36	48	60	0.6	65	156	169	0.3846
77	264	275	0.28	48	90	102	0.4706	91	312	325	0.28
99	440	451	0.2195	60	144	156	0.3846	117	520	533	0.2195
121	660	671	0.1803	72	210	222	0.3243	143	780	793	0.1803
143	924	935	0.1529	84	288	300	0.28	169	1092	1105	0.1529
165	1232	1243	0.1327	96	378	390	0.2462	195	1456	1469	0.1327
187	1584	1595	0.1172	108	480	492	0.2195	221	1872	1885	0.1172
209	1980	1991	0.105	120	594	606	0.198	247	2340	2353	0.105
231	2420	2431	0.095	132	720	732	0.1803	273	2860	2873	0.095

Notes: **DI**: Divison Interval.
For 'Division by ...' series: 1, 2, 8, 9, see Tables 3.1 and 3.2, New Ratio Divisors.
Can be found using Tables 2.1 to 2.4
Can be found using Table 2.5

Table 4.2   **The Multiplying Divisors**

Division by 14		DI:14	7 × (÷2)	Division by 15		DI:30	15 × (÷1)	Division by 16		DI:8	2 × (÷8)
AV1	AV2	Maj. Val.	Sin MiV	AV1	AV2	Maj. Val.	Sin MiV	AV1	AV2	Maj. Val.	Sin MiV
28	21	35	0.6	45	60	75	0.6	24	10	26	0.3846
42	56	70	0.6	75	180	195	0.3846	32	24	40	0.6
56	105	119	0.4706	105	360	375	0.28	40	42	58	0.6897
70	168	182	0.3846	135	600	615	0.2195	48	64	80	0.6
84	245	259	0.3243	165	900	915	0.1803	56	90	106	0.5283
98	336	350	0.28	195	1260	1275	0.1529	64	120	136	0.4706
112	441	455	0.2462	225	1680	1695	0.1327	72	154	170	0.4235
126	560	574	0.2195	255	2160	2175	0.1172	80	192	208	0.3846
140	693	707	0.198	285	2700	2715	0.105	88	234	250	0.352
154	840	854	0.1803	315	3300	3315	0.095	96	280	296	0.3243

Division by 17		DI:34	17 × (÷1)	Division by 19		DI:38	19 × (÷1)	Division by 20		DI:20	10 × (÷2)
AV1	AV2	Maj. Val.	Sin MiV	AV1	AV2	Maj. Val.	Sin MiV	AV1	AV2	Maj. Val.	Sin MiV
51	68	85	0.6	57	76	95	0.6	40	30	50	0.6
85	204	221	0.3846	95	228	247	0.3846	60	80	100	0.6
119	408	425	0.28	133	456	475	0.28	80	150	170	0.4706
153	680	697	0.2195	171	760	779	0.2195	100	240	260	0.3846
187	1020	1037	0.1803	209	1140	1159	0.1803	120	350	370	0.3243
221	1428	1445	0.1529	247	1596	1615	0.1529	140	480	500	0.28
255	1904	1921	0.1327	285	2128	2147	0.1327	160	630	650	0.2462
289	2448	2465	0.1172	323	2736	2755	0.1172	180	800	820	0.2195
323	3060	3077	0.105	361	3420	3439	0.105	200	990	1010	0.198
357	3740	3757	0.095	399	4180	4199	0.095	220	1200	1220	0.1803

Division by 21		DI:42	21 × (÷1)	Division by 22		DI:22	11 × (÷2)	Division by 23		DI:46	23 × (÷1)
AV1	AV2	Maj. Val.	Sin MiV	AV1	AV2	Maj. Val.	Sin MiV	AV1	AV2	Maj. Val.	Sin MiV
63	84	105	0.6	44	33	55	0.6	69	92	115	0.6
105	252	273	0.3846	66	88	110	0.6	115	276	299	0.3846
147	504	525	0.28	88	165	187	0.4706	161	552	575	0.28
189	840	861	0.2195	110	264	286	0.3846	207	920	943	0.2195
231	1260	1281	0.1803	132	385	407	0.3243	253	1380	1403	0.1803
273	1764	1785	0.1529	154	528	550	0.28	299	1932	1955	0.1529
315	2352	2373	0.1327	176	693	715	0.2462	345	2576	2599	0.1327
357	3024	3045	0.1172	198	880	902	0.2195	391	3312	3335	0.1172
399	3780	3801	0.105	220	1089	1111	0.198	437	4140	4163	0.105
441	4620	4641	0.095	242	1320	1342	0.1803	483	5060	5083	0.095

Notes: DI: Divison Interval.

For Division by 18, see Table 3.4, New Ratio Divisors

Can be found using Tables 2.1 to 2.4

Can be found using Table 2.5

Table 4.3

The Multiplying Divisors

Division by 24		DI:12	3 × (÷8)	Division by 26		DI:26	13 × (÷2)	Division by 27		DI:18	3 × (÷9)
AV1	AV2	Maj. Val.	Sin MiV	AV1	AV2	Maj. Val.	Sin MiV	AV1	AV2	Maj. Val.	Sin MiV
36	15	39	0.3846	52	39	65	0.6	45	24	51	0.4706
48	36	60	0.6	78	104	130	0.6	63	60	87	0.6897
60	63	87	0.6897	104	195	221	0.4706	81	108	135	0.6
72	96	120	0.6	130	312	338	0.3846	99	168	195	0.5077
84	135	159	0.5283	156	455	481	0.3243	117	240	267	0.4382
96	180	204	0.4706	182	624	650	0.28	135	324	351	0.3846
108	231	255	0.4235	208	819	845	0.2462	153	420	447	0.3423
120	288	312	0.3846	234	1040	1066	0.2195	171	528	555	0.3081
132	351	375	0.352	260	1287	1313	0.198	189	648	675	0.28
144	420	444	0.3243	286	1560	1586	0.1803	207	780	807	0.2565

Division by 28		DI:28	14 × (÷2)	Division by 29		DI:58	29 × (÷1)	Division by 30		DI:30	15 × (÷2)
AV1	AV2	Maj. Val.	Sin MiV	AV1	AV2	Maj. Val.	Sin MiV	AV1	AV2	Maj. Val.	Sin MiV
56	42	70	0.6	87	116	145	0.6	60	45	75	0.6
84	112	140	0.6	145	348	377	0.3846	90	120	150	0.6
112	210	238	0.4706	203	696	725	0.28	120	225	255	0.4706
140	336	364	0.3846	261	1160	1189	0.2195	150	360	390	0.3846
168	490	518	0.3243	319	1740	1769	0.1803	180	525	555	0.3243
196	672	700	0.28	377	2436	2465	0.1529	210	720	750	0.28
224	882	910	0.2462	435	3248	3277	0.1327	240	945	975	0.2462
252	1120	1148	0.2195	493	4176	4205	0.1172	270	1200	1230	0.2195
280	1386	1414	0.198	551	5220	5249	0.105	300	1485	1515	0.198
308	1680	1708	0.1803	609	6380	6409	0.095	330	1800	1830	0.1803

Division by 31		DI:62	31 × (÷1)	Division by 33		DI:66	33 × (÷1)	Division by 34		DI:34	17 × (÷2)
AV1	AV2	Maj. Val.	Sin MiV	AV1	AV2	Maj. Val.	Sin MiV	AV1	AV2	Maj. Val.	Sin MiV
93	124	155	0.6	99	132	165	0.6	68	51	85	0.6
155	372	403	0.3846	165	396	429	0.3846	102	136	170	0.6
217	744	775	0.28	231	792	825	0.28	136	255	289	0.4706
279	1240	1271	0.2195	297	1320	1353	0.2195	170	408	442	0.3846
341	1860	1891	0.1803	363	1980	2013	0.1803	204	595	629	0.3243
403	2604	2635	0.1529	429	2772	2805	0.1529	238	816	850	0.28
465	3472	3503	0.1327	495	3696	3729	0.1327	272	1071	1105	0.2462
527	4464	4495	0.1172	561	4752	4785	0.1172	306	1360	1394	0.2195
589	5580	5611	0.105	627	5940	5973	0.105	340	1683	1717	0.198
651	6820	6851	0.095	693	7260	7293	0.095	374	2040	2074	0.1803

Notes: DI: Divison Interval.

For 'Division by ...' series: 25, 32, see Tables 3.4 and 3.5, New Ratio Divisors.

Can be found using Tables 2.1 to 2.4

Can be found using Table 2.5

Table 4.4

The Multiplying Divisors

Division by 35		DI:70	35 × (÷1)	Division by 36		DI:12	2 × (÷18)	Division by 37		DI:74	37 × (÷1)
AV1	AV2	Maj. Val.	Sin MiV	AV1	AV2	Maj. Val.	Sin MiV	AV1	AV2	Maj. Val.	Sin MiV
105	140	175	0.6	48	14	50	0.28	111	148	185	0.6
175	420	455	0.3846	60	32	68	0.4706	185	444	481	0.3846
245	840	875	0.28	72	54	90	0.6	259	888	925	0.28
315	1400	1435	0.2195	84	80	116	0.6897	333	1480	1517	0.2195
385	2100	2135	0.1803	96	110	146	0.6575	407	2220	2257	0.1803
455	2940	2975	0.1529	108	144	180	0.6	481	3108	3145	0.1529
525	3920	3955	0.1327	120	182	218	0.5505	555	4144	4181	0.1327
595	5040	5075	0.1172	132	224	260	0.5077	629	5328	5365	0.1172
665	6300	6335	0.105	144	270	306	0.4706	703	6660	6697	0.105
735	7700	7735	0.095	156	320	356	0.4382	777	8140	8177	0.095
Division by 38		DI:38	19 × (÷2)	Division by 39		DI:78	39 × (÷1)	Division by 40		DI:20	5 × (÷8)
AV1	AV2	Maj. Val.	Sin MiV	AV1	AV2	Maj. Val.	Sin MiV	AV1	AV2	Maj. Val.	Sin MiV
76	57	95	0.6	117	156	195	0.6	60	25	65	0.3846
114	152	190	0.6	195	468	507	0.3846	80	60	100	0.6
152	285	323	0.4706	273	936	975	0.28	100	105	145	0.6897
190	456	494	0.3846	351	1560	1599	0.2195	120	160	200	0.6
228	665	703	0.3243	429	2340	2379	0.1803	140	225	265	0.5283
266	912	950	0.28	507	3276	3315	0.1529	160	300	340	0.4706
304	1197	1235	0.2462	585	4368	4407	0.1327	180	385	425	0.4235
342	1520	1558	0.2195	663	5616	5655	0.1172	200	480	520	0.3846
380	1881	1919	0.198	741	7020	7059	0.105	220	585	625	0.352
418	2280	2318	0.1803	819	8580	8619	0.095	240	700	740	0.3243
Division by 41		DI:82	41 × (÷1)	Division by 42		DI:42	21 × (÷2)	Division by 43		DI:86	43 × (÷1)
AV1	AV2	Maj. Val.	Sin MiV	AV1	AV2	Maj. Val.	Sin MiV	AV1	AV2	Maj. Val.	Sin MiV
123	164	205	0.6	84	63	105	0.6	129	172	215	0.6
205	492	533	0.3846	126	168	210	0.6	215	516	559	0.3846
287	984	1025	0.28	168	315	357	0.4706	301	1032	1075	0.28
369	1640	1681	0.2195	210	504	546	0.3846	387	1720	1763	0.2195
451	2460	2501	0.1803	252	735	777	0.3243	473	2580	2623	0.1803
533	3444	3485	0.1529	294	1008	1050	0.28	559	3612	3655	0.1529
615	4592	4633	0.1327	336	1323	1365	0.2462	645	4816	4859	0.1327
697	5904	5945	0.1172	378	1680	1722	0.2195	731	6192	6235	0.1172
779	7380	7421	0.105	420	2079	2121	0.198	817	7740	7783	0.105
861	9020	9061	0.095	462	2520	2562	0.1803	903	9460	9503	0.095

Notes: DI: Divison Interval.

Can be found using Tables 2.1 to 2.4
Can be found using Table 2.5

Table 4.5 — **The Multiplying Divisors**

Division by 44		DI:44	22 × (÷2)	Division by 45		DI:30	5 × (÷9)	Division by 46		DI:46	23 × (÷2)
AV1	AV2	Maj. Val.	Sin MiV	AV1	AV2	Maj. Val.	Sin MiV	AV1	AV2	Maj. Val.	Sin MiV
88	66	110	0.6	75	40	85	0.4706	92	69	115	0.6
132	176	220	0.6	105	100	145	0.6897	138	184	230	0.6
176	330	374	0.4706	135	180	225	0.6	184	345	391	0.4706
220	528	572	0.3846	165	280	325	0.5077	230	552	598	0.3846
264	770	814	0.3243	195	400	445	0.4382	276	805	851	0.3243
308	1056	1100	0.28	225	540	585	0.3846	322	1104	1150	0.28
352	1386	1430	0.2462	255	700	745	0.3423	368	1449	1495	0.2462
396	1760	1804	0.2195	285	880	925	0.3081	414	1840	1886	0.2195
440	2178	2222	0.198	315	1080	1125	0.28	460	2277	2323	0.198
484	2640	2684	0.1803	345	1300	1345	0.2565	506	2760	2806	0.1803

Division by 47		DI:94	47 × (÷1)	Division by 48		DI:24	6 × (÷8)	Division by 51		DI:102	51 × (÷1)
AV1	AV2	Maj. Val.	Sin MiV	AV1	AV2	Maj. Val.	Sin MiV	AV1	AV2	Maj. Val.	Sin MiV
141	188	235	0.6	72	30	78	0.3846	153	204	255	0.6
235	564	611	0.3846	96	72	120	0.6	255	612	663	0.3846
329	1128	1175	0.28	120	126	174	0.6897	357	1224	1275	0.28
423	1880	1927	0.2195	144	192	240	0.6	459	2040	2091	0.2195
517	2820	2867	0.1803	168	270	318	0.5283	561	3060	3111	0.1803
611	3948	3995	0.1529	192	360	408	0.4706	663	4284	4335	0.1529
705	5264	5311	0.1327	216	462	510	0.4235	765	5712	5763	0.1327
799	6768	6815	0.1172	240	576	624	0.3846	867	7344	7395	0.1172
893	8460	8507	0.105	264	702	750	0.352	969	9180	9231	0.105
987	10340	10387	0.095	288	840	888	0.3243	1071	11220	11271	0.095

Division by 52		DI:52	26 × (÷2)	Division by 53		DI:106	53 × (÷1)	Division by 54		DI:18	3 × (÷18)
AV1	AV2	Maj. Val.	Sin MiV	AV1	AV2	Maj. Val.	Sin MiV	AV1	AV2	Maj. Val.	Sin MiV
104	78	130	0.6	159	212	265	0.6	72	21	75	0.28
156	208	260	0.6	265	636	689	0.3846	90	48	102	0.4706
208	390	442	0.4706	371	1272	1325	0.28	108	81	135	0.6
260	624	676	0.3846	477	2120	2173	0.2195	126	120	174	0.6897
312	910	962	0.3243	583	3180	3233	0.1803	144	165	219	0.6575
364	1248	1300	0.28	689	4452	4505	0.1529	162	216	270	0.6
416	1638	1690	0.2462	795	5936	5989	0.1327	180	273	327	0.5505
468	2080	2132	0.2195	901	7632	7685	0.1172	198	336	390	0.5077
520	2574	2626	0.198	1007	9540	9593	0.105	216	405	459	0.4706
572	3120	3172	0.1803	1113	11660	11713	0.095	234	480	534	0.4382

Notes: DI: Divison Interval.

For 'Division by ...' series: 49, 50 see Tables 3.5 and 3.6, New Ratio Divisors.

    Can be found using Tables 2.1 to 2.4

    Can be found using Table 2.5

Table 4.6

The Multiplying Divisors

Division by 55		DI:110	55 × (÷1)	Division by 56		DI:28	7 × (÷8)	Division by 57		DI:114	57 × (÷1)
AV1	AV2	Maj. Val.	Sin MiV	AV1	AV2	Maj. Val.	Sin MiV	AV1	AV2	Maj. Val.	Sin MiV
165	220	275	0.6	84	35	91	0.3846	171	228	285	0.6
275	660	715	0.3846	112	84	140	0.6	285	684	741	0.3846
385	1320	1375	0.28	140	147	203	0.6897	399	1368	1425	0.28
495	2200	2255	0.2195	168	224	280	0.6	513	2280	2337	0.2195
605	3300	3355	0.1803	196	315	371	0.5283	627	3420	3477	0.1803
715	4620	4675	0.1529	224	420	476	0.4706	741	4788	4845	0.1529
825	6160	6215	0.1327	252	539	595	0.4235	855	6384	6441	0.1327
935	7920	7975	0.1172	280	672	728	0.3846	969	8208	8265	0.1172
1045	9900	9955	0.105	308	819	875	0.352	1083	10260	10317	0.105
1155	12100	12155	0.095	336	980	1036	0.3243	1197	12540	12597	0.095
Division by 58		DI:58	29 × (÷2)	Division by 59		DI:118	59 × (÷1)	Division by 60		DI:60	30 × (÷2)
AV1	AV2	Maj. Val.	Sin MiV	AV1	AV2	Maj. Val.	Sin MiV	AV1	AV2	Maj. Val.	Sin MiV
116	87	145	0.6	177	236	295	0.6	120	90	150	0.6
174	232	290	0.6	295	708	767	0.3846	180	240	300	0.6
232	435	493	0.4706	413	1416	1475	0.28	240	450	510	0.4706
290	696	754	0.3846	531	2360	2419	0.2195	300	720	780	0.3846
348	1015	1073	0.3243	649	3540	3599	0.1803	360	1050	1110	0.3243
406	1392	1450	0.28	767	4956	5015	0.1529	420	1440	1500	0.28
464	1827	1885	0.2462	885	6608	6667	0.1327	480	1890	1950	0.2462
522	2320	2378	0.2195	1003	8496	8555	0.1172	540	2400	2460	0.2195
580	2871	2929	0.198	1121	10620	10679	0.105	600	2970	3030	0.198
638	3480	3538	0.1803	1239	12980	13039	0.095	660	3600	3660	0.1803
Division by 61		DI:122	61 × (÷1)	Division by 62		DI:62	31 × (÷2)	Division by 63		DI:42	7 × (÷9)
AV1	AV2	Maj. Val.	Sin MiV	AV1	AV2	Maj. Val.	Sin MiV	AV1	AV2	Maj. Val.	Sin MiV
183	244	305	0.6	124	93	155	0.6	105	56	119	0.4706
305	732	793	0.3846	186	248	310	0.6	147	140	203	0.6897
427	1464	1525	0.28	248	465	527	0.4706	189	252	315	0.6
549	2440	2501	0.2195	310	744	806	0.3846	231	392	455	0.5077
671	3660	3721	0.1803	372	1085	1147	0.3243	273	560	623	0.4382
793	5124	5185	0.1529	434	1488	1550	0.28	315	756	819	0.3846
915	6832	6893	0.1327	496	1953	2015	0.2462	357	980	1043	0.3423
1037	8784	8845	0.1172	558	2480	2542	0.2195	399	1232	1295	0.3081
1159	10980	11041	0.105	620	3069	3131	0.198	441	1512	1575	0.28
1281	13420	13481	0.095	682	3720	3782	0.1803	483	1820	1883	0.2565

Notes: **DI**: Divison Interval.

Can be found using Tables 2.1 to 2.4
Can be found using Table 2.5

Table 4.7 — **The Multiplying Divisors**

Division by 64		DI:16	2 × (÷32)	Division by 65		DI:130	65 × (÷1)	Division by 66		DI:66	33 × (÷2)
AV1	AV2	Maj. Val.	Sin MiV	AV1	AV2	Maj. Val.	Sin MiV	AV1	AV2	Maj. Val.	Sin MiV
80	18	82	0.2195	195	260	325	0.6	132	99	165	0.6
96	40	104	0.3846	325	780	845	0.3846	198	264	330	0.6
112	66	130	0.5077	455	1560	1625	0.28	264	495	561	0.4706
128	96	160	0.6	585	2600	2665	0.2195	330	792	858	0.3846
144	130	194	0.6701	715	3900	3965	0.1803	396	1155	1221	0.3243
160	168	232	0.6897	845	5460	5525	0.1529	462	1584	1650	0.28
176	210	274	0.6423	975	7280	7345	0.1327	528	2079	2145	0.2462
192	256	320	0.6	1105	9360	9425	0.1172	594	2640	2706	0.2195
208	306	370	0.5622	1235	11700	11765	0.105	660	3267	3333	0.198
224	360	424	0.5283	1365	14300	14365	0.095	726	3960	4026	0.1803

Division by 67		DI:134	67 × (÷1)	Division by 68		DI:68	34 × (÷2)	Division by 69		DI:138	69 × (÷1)
AV1	AV2	Maj. Val.	Sin MiV	AV1	AV2	Maj. Val.	Sin MiV	AV1	AV2	Maj. Val.	Sin MiV
201	268	335	0.6	136	102	170	0.6	207	276	345	0.6
335	804	871	0.3846	204	272	340	0.6	345	828	897	0.3846
469	1608	1675	0.28	272	510	578	0.4706	483	1656	1725	0.28
603	2680	2747	0.2195	340	816	884	0.3846	621	2760	2829	0.2195
737	4020	4087	0.1803	408	1190	1258	0.3243	759	4140	4209	0.1803
871	5628	5695	0.1529	476	1632	1700	0.28	897	5796	5865	0.1529
1005	7504	7571	0.1327	544	2142	2210	0.2462	1035	7728	7797	0.1327
1139	9648	9715	0.1172	612	2720	2788	0.2195	1173	9936	10005	0.1172
1273	12060	12127	0.105	680	3366	3434	0.198	1311	12420	12489	0.105
1407	14740	14807	0.095	748	4080	4148	0.1803	1449	15180	15249	0.095

Division by 70		DI:70	35 × (÷2)	Division by 71		DI:142	71 × (÷1)	Division by 73		DI:146	73 × (÷1)
AV1	AV2	Maj. Val.	Sin MiV	AV1	AV2	Maj. Val.	Sin MiV	AV1	AV2	Maj. Val.	Sin MiV
140	105	175	0.6	213	284	355	0.6	219	292	365	0.6
210	280	350	0.6	355	852	923	0.3846	365	876	949	0.3846
280	525	595	0.4706	497	1704	1775	0.28	511	1752	1825	0.28
350	840	910	0.3846	639	2840	2911	0.2195	657	2920	2993	0.2195
420	1225	1295	0.3243	781	4260	4331	0.1803	803	4380	4453	0.1803
490	1680	1750	0.28	923	5964	6035	0.1529	949	6132	6205	0.1529
560	2205	2275	0.2462	1065	7952	8023	0.1327	1095	8176	8249	0.1327
630	2800	2870	0.2195	1207	10224	10295	0.1172	1241	10512	10585	0.1172
700	3465	3535	0.198	1349	12780	12851	0.105	1387	13140	13213	0.105
770	4200	4270	0.1803	1491	15620	15691	0.095	1533	16060	16133	0.095

Notes: DI: Divison Interval.
For **Division by 72**, see Table 3.6, New Ratio Divisors.
[ ] Can be found using Tables 2.1 to 2.4
[ ] Can be found using Table 2.5

Table 4.8

The Multiplying Divisors

Division by 74		DI:74	37 × (÷2)	Division by 75		DI:30	3 × (÷25)	Division by 76		DI:76	38 × (÷2)
AV1	AV2	Maj. Val.	Sin MiV	AV1	AV2	Maj. Val.	Sin MiV	AV1	AV2	Maj. Val.	Sin MiV
148	111	185	0.6	105	36	111	0.3243	152	114	190	0.6
222	296	370	0.6	135	84	159	0.5283	228	304	380	0.6
296	555	629	0.4706	165	144	219	0.6575	304	570	646	0.4706
370	888	962	0.3846	195	216	291	0.6701	380	912	988	0.3846
444	1295	1369	0.3243	225	300	375	0.6	456	1330	1406	0.3243
518	1776	1850	0.28	255	396	471	0.5414	532	1824	1900	0.28
592	2331	2405	0.2462	285	504	579	0.4922	608	2394	2470	0.2462
666	2960	3034	0.2195	315	624	699	0.4506	684	3040	3116	0.2195
740	3663	3737	0.198	345	756	831	0.4152	760	3762	3838	0.198
814	4440	4514	0.1803	375	900	975	0.3846	836	4560	4636	0.1803
Division by 77		DI:144	77 × (÷1)	Division by 78		DI:78	39 × (÷2)	Division by 79		DI:158	79 × (÷1)
AV1	AV2	Maj. Val.	Sin MiV	AV1	AV2	Maj. Val.	Sin MiV	AV1	AV2	Maj. Val.	Sin MiV
231	308	385	0.6	156	117	195	0.6	237	316	395	0.6
385	924	1001	0.3846	234	312	390	0.6	395	948	1027	0.3846
539	1848	1925	0.28	312	585	663	0.4706	553	1896	1975	0.28
693	3080	3157	0.2195	390	936	1014	0.3846	711	3160	3239	0.2195
847	4620	4697	0.1803	468	1365	1443	0.3243	869	4740	4819	0.1803
1001	6468	6545	0.1529	546	1872	1950	0.28	1027	6636	6715	0.1529
1155	8624	8701	0.1327	624	2457	2535	0.2462	1185	8848	8927	0.1327
1309	11088	11165	0.1172	702	3120	3198	0.2195	1343	11376	11455	0.1172
1463	13860	13937	0.105	780	3861	3939	0.198	1501	14220	14299	0.105
1617	16940	17017	0.095	858	4680	4758	0.1803	1659	17380	17459	0.095
Division by 80		DI:40	10 × (÷8)	Division by 82		DI:82	41 × (÷2)	Division by 83		DI:166	83 × (÷1)
AV1	AV2	Maj. Val.	Sin MiV	AV1	AV2	Maj. Val.	Sin MiV	AV1	AV2	Maj. Val.	Sin MiV
120	50	130	0.3846	164	123	205	0.6	249	332	415	0.6
160	120	200	0.6	246	328	410	0.6	415	996	1079	0.3846
200	210	290	0.6897	328	615	697	0.4706	581	1992	2075	0.28
240	320	400	0.6	410	984	1066	0.3846	747	3320	3403	0.2195
280	450	530	0.5283	492	1435	1517	0.3243	913	4980	5063	0.1803
320	600	680	0.4706	574	1968	2050	0.28	1079	6972	7055	0.1529
360	770	850	0.4235	656	2583	2665	0.2462	1245	9296	9379	0.1327
400	960	1040	0.3846	738	3280	3362	0.2195	1411	11952	12035	0.1172
440	1170	1250	0.352	820	4059	4141	0.198	1577	14940	15023	0.105
480	1400	1480	0.3243	902	4920	5002	0.1803	1743	18260	18343	0.095

Notes: DI: Divison Interval.

For Division by 81, see Table 3.7, New Ratio Divisors.

     Can be found using Tables 2.1 to 2.4

     Can be found using Table 2.5

Table 5	Reverse Triples arising in Divisors 1 to 82, and Divisor 98										
Normal Generated		Reverse Generated		Normal Generated		Reverse Generated		Normal Generated		Reverse Generated	
Divisor	Triple	Divisor	Rev. Tr.	Divisor	Triple	Divisor	Rev. Tr.	Divisor	Triple	Divisor	Rev. Tr.
1	3,4,5	2	4,3,5	8	40,96,104	64	96,40,104	25	75,100,125	50	100,75,125
1	5,12,13	8	12,5,13	8	44,117,125	81	117,44,125	25	85,132,157	72	132,85,157
1	7,24,25	18	24,7,25	9	27,36,45	18	36,27,45	25	95,168,193	98	168,95,193
1	9,40,41	32	40,9,41	9	33,56,65	32	56,33,65	26	78,104,130	52	104,78,130
1	11,60,61	50	60,11,61	9	39,80,89	50	80,39,89	27	81,108,135	54	108,81,135
1	13,84,85	72	84,13,85	9	45,108,117	72	108,45,117	28	84,112,140	56	112,84,140
1	15,112,113	98	112,15,113	9	51,140,149	98	140,51,149	29	87,116,145	58	116,87,145
2	6,8,10	4	8,6,10	10	30,40,50	20	40,30,50	30	90,120,150	60	120,90,150
2	8,15,17	9	15,8,17	10	40,75,85	45	75,40,85	31	93,124,155	62	124,93,155
2	10,24,26	16	24,10,26	10	50,120,130	80	120,50,130	32	80,84,116	36	84,80,116
2	12,35,37	25	35,12,37	11	33,44,55	22	44,33,55	32	88,105,137	49	105,88,137
2	14,48,50	36	48,14,50	12	36,48,60	24	48,36,60	32	96,128,160	64	128,96,160
2	16,63,65	49	63,16,65	12	48,90,102	54	90,48,102	32	104,153,185	81	153,104,185
2	18,80,82	64	80,18,82	13	39,52,65	26	52,39,65	33	99,132,165	66	132,99,165
2	20,99,101	81	99,20,101	14	42,56,70	28	56,42,70	34	102,136,170	68	136,102,170
3	9,12,15	6	12,9,15	14	56,105,119	63	105,56,119	35	105,140,175	70	140,105,175
3	15,36,39	24	36,15,39	15	45,60,75	30	60,45,75	36	96,110,146	50	110,96,146
3	21,72,75	54	72,21,75	16	40,42,58	18	42,40,58	36	108,144,180	72	144,108,180
4	12,16,20	8	16,12,20	16	48,64,80	32	64,48,80	36	120,182,218	98	182,120,218
4	16,30,34	18	30,16,34	16	56,90,106	50	90,56,106	37	111,148,185	74	148,111,185
4	20,48,52	32	48,20,52	16	64,120,136	72	120,64,136	38	114,152,190	76	152,114,190
4	24,70,74	50	70,24,74	16	72,154,170	98	154,72,170	39	117,156,195	78	156,117,195
4	28,96,100	72	96,28,100	17	51,68,85	34	68,51,85	40	100,105,145	45	105,100,145
4	32,126,130	98	126,32,130	18	48,55,73	25	55,48,73	40	120,160,200	80	160,120,200
5	15,20,25	10	20,15,25	18	54,72,90	36	72,54,90	41	123,164,205	82	164,123,205
5	25,60,65	40	60,25,65	18	60,91,109	49	91,60,109	48	120,126,174	54	126,120,174
6	18,24,30	12	24,18,30	18	66,112,130	64	112,66,130	49	119,120,169	50	120,119,169
6	24,45,51	27	45,24,51	18	72,135,153	81	135,72,153	49	133,156,205	72	156,133,205
6	30,72,78	48	72,30,78	19	57,76,95	38	76,57,95	49	147,196,245	98	196,147,245
6	36,105,111	75	105,36,111	20	60,80,100	40	80,60,100	50	130,144,194	64	144,130,194
7	21,28,35	14	28,21,35	21	63,84,105	42	84,63,105	50	140,171,221	81	171,140,221
7	35,84,91	56	84,35,91	22	66,88,110	44	88,66,110	54	144,165,219	75	165,144,219
8	20,21,29	9	21,20,29	23	69,92,115	46	92,69,115	56	140,147,203	63	147,140,203
8	24,32,40	16	32,24,40	24	60,63,87	27	63,60,87	64	160,168,232	72	168,160,232
8	28,45,53	25	45,28,53	24	72,96,120	48	96,72,120	64	176,210,274	98	210,176,274
8	32,60,68	36	60,32,68	24	84,135,159	75	135,84,159	72	180,189,261	81	189,180,261
8	36,77,85	49	77,36,85	25	65,72,97	32	72,65,97	81	207,224,305	98	224,207,305

Shown in Tables 2.1 to 2.4
Can be found using Tables 2.1 to 2.4

BONES 2: ISOLATING PRIMES

<u>Main Index:</u> 3 to 109999, odd numbers only. (This makes use of 2 as first Prime Identifier)

<u>Prime Identifiers required</u>: Given by: $\sqrt{109999}$, Prime 3 to Prime 331 (Total: 66)

<u>Summation Algorithm:</u>
     Using Prime Identifiers 3 to 43, Total: 13 identifiers. Each progresses on step value: 2×Prime. Each starts at: PI².

<u>Multiplier Algorithm:</u> Given by: $\sqrt[3]{109999} = 47.91...$
     Prime Identifiers used: 47 to 331
     'Primes only' multiplier: Range of primes 47 to 2339. (333 numbers required.) Note: 47 is just below $\sqrt[3]{}$(Test Limit), therefore non-prime multiplier value 2209 (47²) is required. Prime Identifier 47 was the only identifier to use this multiplier value and the prime multipliers greater than 2209.
     Each Prime Identifier in this algorithm started at its own value in the multiplier index and progressed through the prime multiplier values to the highest product before test limit.

<u>Primes Identified:</u> Those numbers unidentified by either algorithm.

<u>Summary of Results</u>
     1. Total odd numbers, 3 to 109999:   54999
     2. Total primes isolated: 10453
     3. Total non-primes in the range (1-2)   44546
     4. Total marks, Multiplier Algorithm (4.7%)   5184
     5. Total numbers marked by Summation Algorithm(3-4)   39362
     (6. Total marks placed by Sum. Alg.: 62586)
     (7. Effective marking by Sum. Alg.: 62.9%)

<u>Isolating vis a vis the Sieve</u>
     The sieving method does not produce any second marking (see 6 & 7 above) because it crosses out numbers as they are identified as non-prime. If this is taken literally, as obliteration, then the progress of identifiers is hampered by having to identify what is factorised by the later identifiers. In other words, one ends up prime testing. Having a continuing Main Index allows summation progression, but produces a high rate of second marking.

BONES 3: THE SQUARE, INCREASING IT AND GROWING IT

The method of increasing the square is well known. The diagram below shows the outline of what is done, using a square of side x and increasing it by one unit, which in total would become $(x + 1)^2$.

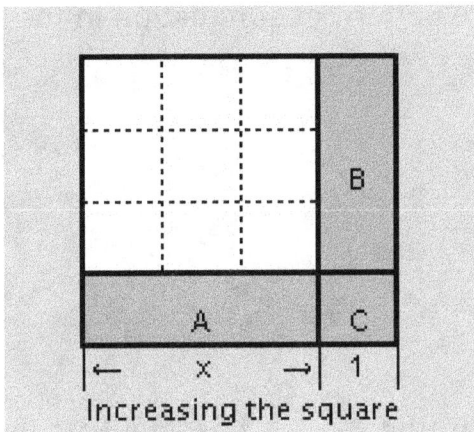

Increasing the square

The increase is given by the three areas A, B, and C. Areas A and B measure x × 1 and area C measures 1 × 1. So the areas of increase could be added together and mathematically expressed as:

$$2(1 \times x) + 1^2$$

Obvoiusly that is going to come out as 2x + 1. Consider what would happen if the mathematician wishes to handle an increase of the square of let's say, 277. It's worth remembering a point about squareness and its increase. It is very easy to show it as a diagram, translating the squareness into surface area. However, the squaring of a value can be seen as an abstract in much the same way that we add 2 and 2 to get 4. We don't have to say 2 of what. So this supposed increase of 277 could be an abstract concept. As with 2x+1, the objective is to state the increase for all values of x. The precisely correct expression shown above would now read:

$$2(277 \times x) + 277^2$$

which would be: $554x + 277^2$

and here the mathematician comes up against the daftness of his 'linear' label. The expression is still in the same family as 2x + 1, so it can't show a 'squared' sign so he writes: 554x + 76729. Never mind that 554x is an entity of area, the same as the 2x in the 2x + 1.

Growing the square is a different matter. For this we need to look at the full statement of the x^2 series, complete with both lines of difference. The diagram below shows the series with the diagonal lines indicating the lattice pattern of summation in the series.

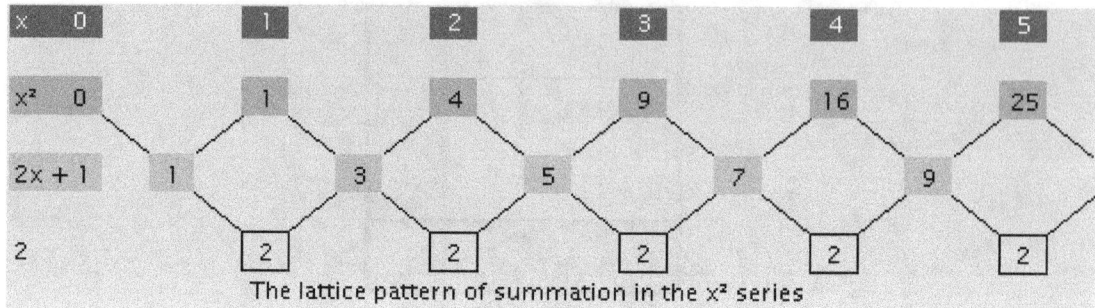

The lattice pattern of summation in the x^2 series

Each value from the 2x + 1 line, the first line of difference, works twice. It adds to the previous square to form the next square. Then it picks up the next 2 from the second line of difference to add to its own value to then form the next first difference. It is not as simple as one might suppose. More to the point it is asking for something continuous to be done – 'Take the previous increase and add 2'. A continuous entity that retains the essential of squareness is the square spiral.

73	74	75	76	77	78	79	80	81
72	43	44	45	46	47	48	49	50
71	42	21	22	23	24	25	26	51
70	41	20	7	8	9	10	27	52
69	40	19	6	1	2	11	28	53
68	39	18	5	4	3	12	29	54
67	38	17	16	15	14	13	30	55
66	37	36	35	34	33	32	31	56
65	64	63	62	61	60	59	58	57

The square spiral - growing the square

The square spiral is well known to mathematics. I have never seen it represented as the imperative of summation in the x^2 series.

Obviously the next question would be, "Can a similar kind of growth, as distinct from increase, be seen in the x^3 series and beyond?"

It is not apparent to me. Below are the first stages of the x^3 series. It may be the case that a distinction between increase and growth is not possible. However, the two elements seen in the x^2 series are present in this series. Using only the first line of difference, a simple statement of increase can be made. The lattice pattern of summation is there and this may indicate that something different can be done.

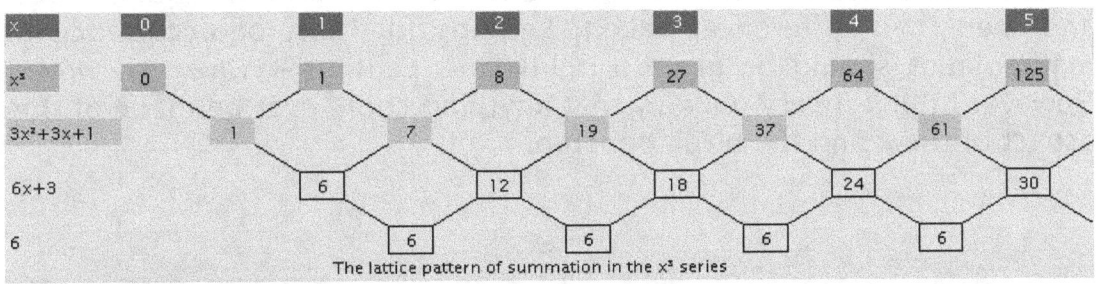

The lattice pattern of summation in the x^3 series

It can be argued that these differences are simply that, and they mean nothing. If that argument could have been made in the past it can no longer. The Triples Generator has put paid to that. Perhaps even the seemingly small item of the square spiral can be weighed in too.

Another aspect of x^n would be to look at a specific value of x raised through the successive values of n to see whether the differences there might shed any light. The table below shows x value of 2 raised through the sequence of n from 0 to 6.

n	0	1	2	3	4	5	6
x=2	1	2	4	8	16	32	64
Diff.	1	2	4	8	16	32	64

and repeat, towards infinity

The repeated line of differences only happens when 2 is raised through the sequence of n. In fact it appears that there is an 'increase' in successive lines of difference. It is given by (x–1), acting as a multiplier on each line of difference. Obviously, in the case of $x = 2$ that will result in a repeat sequence.

n	0		1		2		3		4		5		6
x=3	1		3		9		27		81		243		729
Diff(1)		2		6		18		54		162		486	
Diff(2)			4		12		36		108		324		972

and ×2 per line of difference, towards infinity

The start of the x = 3 series is shown above. Here the increase of lines of differences is 2×, still a multiplier of (x−1). When x=4 the multiplier for successive lines of difference is 3. This pattern appears to continue as successive values of x are raised through the sequence of n and then their differences taken. Perhaps this sort of occurrence in mathematics should be given a politician's caution – *Never say never.* However I think it very unlikely that anything could ever be made of this aspect of the x and the n relationship.

EPILOGUE

JOE AND FRED, AND SHEILA

Fred Martin rang Joe the day he realised that Joe's idea was more than it had first appeared. They met that evening in a pub near the university. Joe was pleased to know that the idea was right after all and, as Fred started to work his way round to the question of attribution, Joe sensed what was coming and saved Fred the bother.

"Just get on with it and stop worrying. It's more than enough to know I had a good idea. You've got all the work to do. I just lobbed the pebble in the pond."

After that the two men got on well together. They would meet occaisionally and the friendship became a firm one.

Fred's paper had done his career some good, though he stayed in the same university. He liked the city and he liked the work he did in mathematics. Over the years he took on some departmental responsibilities. Like most people who have a special talent, he didn't like administrative work much but it had to be done.

The teaching side has always interested and stimulated Fred. He likes those times when someone suddenly sees the point, the identification Fred has been working towards – it's always an off-the-wall moment, a real kick.

Joe gave up working on the bins a couple of years after meeting Fred. He reckons it's a young man's game. He'd done some tough jobs, but for sustained strain on the back and the knees, the bins, at least the bag system, took some beating.

He found a job working for a printing and bookbinding company. By then he was thirtysomething. He'd done the wild young man thing starting from fifteen but he knew in his heart it had been a waste. He would have liked to learn a skilled trade. When Joe started the apprenticeship system was still working and nobody would even think of him for learning a skill – he was just too old.

He worked as a general dogsbody for the company, driving, cleaning and assisting the printers. It was the bookbinding he liked. He wasn't above admitting that he envied the two young apprentices in the bookbinding room.

Times change. The apprentices were close mates. One of them had an uncle on the oil rigs and they both left to work there. Joe saw the young men as doing what he'd done, only taking longer to get started. He wondered if it would work out better for them.

Albert, the specialist bookbinder, went to see the boss. Albert had been with the company all his life. He'd started in the days when there were still gentlemen who would have everything in their library that wasn't leatherbound, rebound in leather. As a result of that experience, Albert became the finest craftsman the firm had ever employed. His point to the boss: *"I know Joe would like to learn. I reckon he'll learn quick. Put Joe with me. Take on a new van driver."*

The old apprenticeship rules had gone, a centuries old tradition vanished in a couple of years. There was a steady stream of work, all individual pieces. Joe did learn quickly and he got on well with Albert. Sometimes they would meet up with Fred for a drink after work.

Joe set up a workshop in his flat, just for his own interest, and for Fred's. Some of the most unlikely books have been given the leather treatment. Early pieces were two *Perishers* annuals for Fred and Joe followed through with his *Giles* annuals; some Asimov, Arthur C. Clarke and Ray Bradbury. They both like SF. Fred has found he likes this hobby. He is better with his hands than he thought he would be.

Two years ago Fred brought in *Who Runs This Place?* by Anthony Sampson. Joe thought that a strange choice, until he read it. They discussed doing the highly entertaining flip of Sampson's work, the *Yes, Minister* and *Yes, Prime Minister* series. In the end they agreed that for these the publisher had got the bindings, well, the dust jacket really, just right. It seemed to speak to both the entertainment and the pretension the books so effectively poked. But the Sampson, that was a different matter. For both men the only criterion was: do I think the book is worth it?

Last year Fred finally got round to reading Brian Butterworth's *The Mathematical Brain*. In a way it was where the two men had started all those years ago. So now Joe and Fred are working on two copies.

I cannot bring myself to write the pun.

Sheila Bigouad logged on from her retirement home in Tasmania and saw a very bright light. It was a new light and shone straight across the world of mathematics.

She had retired more than thirty years previously and was now ninety-three. Mathematics was still fascinating, but the conjecture left her with very mixed feelings. It had started towards the end of the Second World War and had involved a lot of number crunching. In those days it meant slide rule and log tables. Her fiancé, Peter Lewis, was away for long periods in the Australian navy. She was steadily working through until he returned home.

He had a short leave and all these years on she had not known why they did not marry then, but they didn't. Both were so certain it would happen right for them anyway that it had seemed not to matter. Three months later Peter was dead, killed in an on-board accident, the kind of thing that happens when men and equipment are stressed too far. There was no one to blame, and in the end that had given Sheila some relief. But the relief of not needing to pick up hatred and then having to fight to put it down again had taken a long time to come. In the meantime the grief, loss, and regret looked like they would drown her.

For two months her work had stopped. Suddenly she started again and her friends wondered if she was becoming manic. Sheila knew it might be the case but work was the only antidote for her deep grief. The work that was the basis for the conjecture was finished in 1946. She put it away in the drawer and it stayed there for ten years.

Through those years she worked and she recovered, from the grief and regret, and she recovered her real personality, a cheerful, warm and friendly person. She had always enjoyed the company of others and she would help anyone who needed her. She had submitted the conjecture in 1956. Few knew of the tragic associations it held for her.

The light had come from a hole in a wall. The wall had been known for some time. It had a number of small holes either side of the solid centre. An early insight suggested that the solid central region was, in reality, a doorway; it could be removed as a complete unit. The problem was cracking it free. Many methods had been tried and none worked. Someone had been doing some fiddling around with one of the holes out along the wall. A small mathematical device was turned in the gap and the door came free. The locking mechanism was in the wall, not the door.

In fact there were a number of locks, but any one of them would free the door. A great discovery. Much work to be done. Surveying the landscape beyond the wall. Fit of door to wall. Why are there locking bars running to these holes, but not the others? Someone suggested trying to patent locks in walls, but it's as old as the hills. What nobody noticed was the light coming through the doorway, nor its direction.

No one that is except for Sheila. The conjecture had stayed on the back burner for most of its life. But what Sheila saw was a light that shone straight into the room of The Bigouad Conjecture. She made the proof by the means she had always used for doing mathematics – a pad of plain paper and a 2H pencil. She left it and checked the next day. Only then did it go onto her computer and was downloaded to the journal. A while later it was published. The Bigouad Conjecture became The Bigouad Theorem.

For Sheila it was an ending. No one now alive could know its true significance for her. Peter had been her spiritual companion all her life. This was the completion of their spritual love and a preparation for whatever it may be that is to come. She died peacefully in her sleep two years later.

ABOUT THE AUTHOR

I do not have someone to write a profoundsound blurb about me, so it will have to be first person.

I was born in 1946, in Coventry, England. I have many interests of which art, in particular sculpture, politics, and the more simple elements of mathematics are a few.

My life thus far has had very little of any real note. I've made a number of sculptures, almost all of them wood carvings, and sold a few; a couple of inventions, none taken up. It's been that sort of a life.

On credibility tokens, I have a degree in politics. If you have read this book, then ask yourself, *"Does it make sense?"* If it does, then it is the sense that matters, not who wrote it.

I am interested in history but I am far more concerned about where humans are going than where they have been.

My wife and I live with our five cats in Birmingham, England.